Lux, David Stephan.

Patronage and royal science in seventeenth-century France.

$28.50

DATE			

Patronage and Royal Science in Seventeenth-Century France

Patronage and Royal Science in Seventeenth-Century France

The Académie de Physique in Caen

David S. Lux

Cornell University Press

ITHACA AND LONDON

THIS BOOK IS PUBLISHED WITH THE AID OF A GRANT FROM THE
VIRGINIA POLYTECHNIC INSTITUTE AND STATE UNIVERSITY.

First published 1989 by Cornell University Press.

International Standard Book Number 0-8014-2334-1
Library of Congress Catalog Card Number 89-1002

Printed in the United States of America

The paper in this book is acid-free and meets the guidelines for permanence and durability of the Committee on Production Guidelines for Book Longevity of the Council on Library Resources.

Dedicated to those who deserve it most,

Sharon, Jessica, Hillary, and Madeleine

Contents

Preface ix
Abbreviations xv

Introduction 1
1 The Academy's Origins 8
2 The *Assemblée* Becomes an Academy 29
3 The Dynamics of a Scientific Organization 57
4 The Royal Incorporation 81
5 The Royal Academy of Sciences in Caen, 1668–1669 121
6 The Royal Academy of Sciences in Caen, 1670–1672 140
7 Royal Administration, Patronage, and Science 164
8 Conclusion 180

Bibliography 183

Index 197

Preface

A great many of my ideas about provincial savants in seventeenth-century France have changed in the years since I first investigated them, but nothing has ever shaken my initial sense of wonder at how the members of the Académie de Physique de Caen permitted their loyalty to their patron to override the opportunity to move their collective activity, their organization, onto the grander stage of royal patronage, with its considerable honors, status, rights, and privileges. Their story forms a crucial chapter in the history of science and of French learned societies.

Spanning just the decade between 1662 and 1672, the Académie de Physique's active life was extremely short. Yet the brevity of its existence should not imply that the academy's story merely offers "a slice of life" from seventeenth-century science. Change and novelty constitute the overriding theme in this organization's history. The 1660s saw dramatic alterations in France's intellectual, social, political, and economic life. Provincial as it was, the Académie de Physique reflected those changes. Alterations in the organization and practices of the larger European scientific community also did much to shape the academy's history. Even in the attempt to establish and formalize connections to Paris and the monarchy, the academy's story constitutes a break in the regional history of Normandy's notorious fractiousness. When we consider the depth of change found in the academy's history, it is helpful to reflect that in their youth (during the late 1630s) these Caennais of the 1660s had witnessed the brutal repression royal troops visited on the *Nu-pieds*. Their back-and-forth bickerings over royal incorporation took bizarre form at times, but these were men who found real meaning in the issues of local loyalty and royal splendor.

The Académie de Physique's history addresses issues in French history, but it finds its special meaning in the history of science. The habits of mind, the procedures, the ways of expressing ideas among these academicians were far different from anything that familiarity with modern science might lead us to expect. Indeed, the academy's significance in the history of science rests as much on its members' procedures and mental processes as on any other issue. The major theme of this book is the developing relationships between patronage and royal incorporation; its most important subtext concerns the location of the practice of science within the seventeenth century's hierarchy of intellectual activity. As far as possible, I have tried to formulate this investigation in ways that will create empathy for premodern scientific positions. This approach results not from any antiquarian attempt to recreate the past but rather from my firm belief that historians of science consistently underestimate the sophistication required to formulate both modern and early-modern procedures in laboratory science.

The Académie de Physique's rationale lay in its organizational mission to establish the laboratory as the new locus for scientific practice. Such a task required the formulation of both the specifics of procedure and the organizational form that gives meaning to laboratory science. This was no mean feat, and over the course of the academy's ten-year history we find evidence of numerous conceptual shifts and accompanying changes in forms of expression. Many of these changes are analyzed here, but to try to tackle all such issues directly would bog the entire enterprise down in pedantry. Thus certain terms are simply used in their original French forms without explicit definition or detailed explanation. The most obvious of such terms is the name of the academy itself. There is no satisfactory English translation of "Académie de Physique." "Academy of Physics" would be hopelessly misleading in its connotations. "Academy of Natural Sciences" would be almost accurate for some uses but would obscure needlessly the academy's real divisions and its struggle to come to a tacit definition of purpose.

Many other terms receive the same treatment in this book. *Assemblée, fidèle, chef, gratification, gloire, curieux, expérience*—these are all words that can be given rough equivalents in modern English, but the use of such equivalents would imply both that unitary translations are possible and that these terms stood frozen in meaning during the 1660s. Nothing could be more misleading, since much of the academy's history involves attempts to hammer out the meanings and purposes of such terms as *chef, curiosité,* and even *académie.* Of course, overinsistence on linguistic purity can become cumbersome, most notably in regard to the term "patron." One finds elements of that English concept in *fondateur, instituteur, protecteur,* and *chef,* but neither *le patron* nor *le patronage* appears in any document related to the Académie de Physique. One last note on terminology: my reasons for using the archaic term Académie Royale to denote what is now most often called the Paris Academy of Sciences have to do with contemporary perceptions. For these Caennais and their associates, *Royale* was the operative word.

In dealing with transcriptions of archival sources I have tried to follow the pragmatic principle that simplicity yields the greatest benefit to scholarship. I have provided literal renderings of orthography while editing capitalization and punctuation. My reason for this dual approach rests on my belief that seventeenth-century French orthography carries meanings that warrant preservation—especially as those meanings found expression in André Graindorge's rich language. Balanced against that desire to preserve expression, however, the requirement for clarity in presentation has dictated standardization in capitalization and puntuation.

In citing documents, I have held once again to a pragmatic approach. In all references to documents, I cite the earliest form available to me; I have avoided references to translations and heavily edited texts, with their inevitable errors and omissions. Transcriptions of documents (French and Latin) appear in the notes to this book only where the ability to understand, interpret, or judge the book's narrative is furthered through immediate access to a text. Of course, I realize scholars can disagree significantly on the conditions under which such a restriction applies. Thus I offer this notice: Readers seeking access to archival documents will be served most easily and reliably by microfilm orders placed with the libraries and archives mentioned in the Note on Archival Sources in the Bibliography. If that approach should prove impractical, edited transcriptions of the relevant passages in all references to Pierre-Daniel Huet's papers (including the Graindorge letters) appear in the works by

Léon G. Pelissier, Harcourt Brown, Léon Tolmer, Katherine Brennan, and David Lux cited in the Bibliography.

In the course of this book's development I have incurred many debts that I can only acknowledge, with no hope of ever repaying them in full. Some of these debts are institutional; many are personal. A National Science Foundation Graduate Fellowship supported me for the three years in which the first research took shape. The University of Michigan's Horace Rackham School of Graduate Studies provided a fellowship that allowed the first archival research. The Virginia Polytechnic Institute and State University Humanities Program then provided a summer fellowship that permitted further research. The History Department at Virginia Tech has consistently granted reduced teaching loads and continuing funds that have enabled me to carry forward my research, writing, and revisions. Finally, many librarians and archivists in France, Italy, England, and the United States have helped me along the way. I am particularly grateful to the director and staff of the Biblioteca Medicea-Laurenziana in Florence for arranging access to the Huet papers at a time when the library was closed for restoration of its magnificent reading room.

Friends, teachers, and colleagues have all provided essential aid during my work on the Académie de Physique's history. Harold J. Cook and Hilton Root helped me plant my feet in the history of science and French history, respectively: the two areas of history fundamental to this work. Many of my teachers at the University of Michigan will find their ideas given new forms here. I only hope they recognize those ideas and make allowances for any untoward vulgarizations. Some of my most basic historical concepts came from Charles Trinkaus, who helped me to understand how ideas lead to action; from David D. Bien, who introduced me to the subtleties and power of French history; and from Marvin Becker, whose lectures and questioning revealed both the importance of historical change and the ways in which one century's aberrant social movements can become the next century's institutional history. I am also grateful for the contribution of the late James A. Vann III to the development of my historical consciousness. Much of my institutional analysis and a great deal of my expository style derive from his teachings. I am especially indebted to Nicholas H. Steneck, who not only taught me a great deal about the history of science but also encouraged me to ground that history in the broader currents of the past.

Other friends, colleagues, and professionals have given immeasurable assistance to the progress of this work. Roger Ariew, Peter Barker, and Joseph C. Pitt have greatly expanded my understanding of the philosophy of science. Charles J. Dudley, Ellsworth R. Fuhrman, and William E. Snizek have offered timely commentaries that led me to a greater understanding of sociological theory. Henry H. Bauer has been a continuing source of stimulation for my interest in disciplinary specialization. Mordechai Feingold and Cynthia Bouton have offered important help with the history of early-modern Europe. Thomas Dunlap, Harold C. Livesay, Neil Larry Shumsky, and Peter Wallenstein are American historians who have proved themselves invaluable colleagues in their willingness to read and comment on anything from sentences and paragraphs to chapters and manuscript drafts. I also offer special thanks to the editorial staff of Cornell University Press. John Ackerman, Barbara Salazar, and Lois Krieger have provided immeasurable assistance in completing the final revisions and editing necessary to transform this work from manuscript to book. I am grateful to all these friends, teachers, colleagues, and professionals.

Any academic family also deserves special thanks. Sharon does not type manuscripts, but she has proved herself extraordinarily perceptive in spotting weak arguments and of incalculable support in keeping me from feeling stupid about such gaffes. Our children have also contributed to this work. Jessica was born during the writing of an early draft of Chapter 1. Hillary arrived in the lull between completion of an early version and research for revisions. Madeleine has come into our midst during revisions of the work. All three bear its mark, and they deserve my thanks.

DAVID S. LUX

Blacksburg, Virginia

Abbreviations

AdS, Reg	Académie des Sciences, Paris: Archives, Registre des procès-verbaux des séances
AS	*Annals of Science*
BL, A 1866	Biblioteca Medicea-Laurenziana, Florence: Ashburnham 1866
BN	Bibliothèque Nationale, Paris
CHO	Henry Oldenburg, *The Correspondence of Henry Oldenburg,* ed. A. Rupert Hall and Marie Boas Hall, 13 vols. (Madison: University of Wisconsin Press, 1965–1973 [vols. 1–9]; London: Mansell, 1975–1977 [vols. 10–11]; London: Taylor & Francis, 1986 [vols. 12–13])
HARS	*Histoire de l'Académie Royale des Sciences (1666–1699),* 2 vols. (Paris, 1733)
LIMC	Pierre Clément, ed., *Lettres, instructions, et mémoires de Colbert,* 8 vols. (Paris, 1861–1882)
MANC	*Mémoires de l'Académie Nationale des Sciences, Arts, et Belles-Lettres de Caen*
RHS	*Revue d'Histoire des Sciences et de Leurs Applications*
VDLI	Léon Tolmer, "Vingt-deux Lettres inédites d'André Graindorge à P.-D. Huet," *MANC,* n.s. 12 (1942): 245–337

Introduction

Two events mark the Académie de Physique de Caen as an important subject for historical study.[1] The first occurred in late 1667, when Louis XIV gave the organization his personal *approbation,* or "recognition," making it a royal academy of sciences. The second came in late 1672, when the academy closed after long and bitter disputes over both its "inactivity" and its "waste" of royal funds. Those two events make the academy's story a significant chapter in the history of early-modern scientific institutions. More specifically, since the organization's traceable life ran from 1662 to 1672, its history falls squarely within the period Bernard de Fontenelle cited as inaugurating the "New Age of Academies."[2]

[1]Several works treat various aspects (or periods) of the Académie de Physique's history. They include Harcourt Brown, *Scientific Organizations in Seventeenth-Century France* (New York, 1967 [1934]), pp. 216–230, and "L'Académie de Physique de Caen (1666–1675) d'après les lettres d'André de Graindorge," *MANC,* n.s. 9 (1939): 117–208; Léon Tolmer, "Vingt-deux Lettres inédites d'André de Graindorge à P.-D. Huet," *MANC,* n.s. 12 (1942): 245–337, and *Pierre-Daniel Huet, 1630–1721: Humaniste-physicien* (Bayeux, n.d. [1949]), pp. 271–409; Katherine Stern Brennan, "Culture and Dependencies: The Society of the Men of Letters of Caen from 1652 to 1705 (Ph.D. diss., Johns Hopkins University, 1981), pp. 141–181.

[2]*HARS,* 1:5. Any list of the most basic historical works recently published on the complex of social and institutional changes that directly affected European science, the practice of science, and scientific education following the mid–seventeenth century would include L. W. B. Brockliss, *French Higher Education in the Seventeenth and Eighteenth Centuries: A Cultural History* (Oxford: Clarendon, 1987); Harold J. Cook, *The Decline of the Old Medical Regime in Stuart London* (Ithaca, N.Y., 1986); Mordechai Feingold, *The Mathematicians' Apprenticeship: Science, Universities and Society in England,*

The Académie de Physique's story adds a new dimension to the characterizations historians usually offer to describe Fontenelle's "New Age of Academies." The 1650s through the 1670s did indeed form a period of dramatic reorganization in the social forms governing European science. Important changes occurred rapidly during those three decades. As appears so clearly in the English case, for example, ideological and philosophical issues significantly affected the social reorganization of science during the 1660s. Likewise, as in the case of Parisian science, areas of activity became increasingly specialized in the 1660s and 1670s.

The Académie de Physique's history certainly reflects such larger changes. Yet it also adds something entirely new to our understanding of the organizational changes that affected science in the mid–seventeenth century. We see a state-funded institution cease to exist five years after it received its royal charter. In 1672 the Académie de Physique simply disappeared. The mere fact that such a thing could happen in the France of the 1660s and 1670s adds a new dimension to our understanding of the history of scientific organizations.

The Académie de Physique's history is a study in failure; the pages that follow amply document that fact. This book, however, is not primarily about failure. My purposes go far beyond an attempt to demonstrate something "wrong" with the Académie de Physique. I do not expose a dark, seamy underbelly to seventeenth-century science. Quite the opposite: the academy's history points up just how difficult it was to establish a viable scientific institution in seventeenth-century France. In seeking to understand its failure we can hope to learn what it took to create a successful scientific organization in the 1660s.

The Académie de Physique was an extraordinary organization. By the standards of the day, it boasted an extremely talented membership. It first enjoyed the support of a sympathetic private patron and

1560–1640 (Cambridge, Eng., 1984); Roger Hahn, *The Anatomy of a Scientific Institution: The Paris Academy of Sciences, 1666–1803* (Berkeley, 1971); Michael Hunter, *Science and Society in Restoration England* (Cambridge, Eng., 1981); James R. Jacob and Margaret Jacob, "The Anglican Origins of Modern Science," *Isis* 71 (1980): 251–267; W. E. Knowles Middleton, *The Experimenters: A Study of the Accademia del Cimento* (Baltimore, 1971); Margery Purver, *The Royal Society: Concept and Creation* (Cambridge, Mass., 1967); Barbara J. Shapiro, *Probability and Certainty in Seventeenth-Century England* (Princeton, 1983); Howard M. Solomon, *Public Welfare, Science, and Propaganda in Seventeenth-Century France* (Princeton, 1972); and Charles Webster, *The Great Instauration: Science, Medicine, and Reform, 1626–1660* (New York, 1975).

later received considerable benefits from Louis XIV and his minister Colbert. Even the Académie Royale des Sciences in Paris extended offers of technical support and scientific cooperation to the group in Caen, the only provincial adjunct to the early Académie Royale.[3] Moreover, the Académie de Physique managed to develop a significant research potential, and such respected contemporaries as Henry Oldenburg and Robert Boyle watched its progress with interest. Between 1662 and 1666 its program moved from the world of traditional natural philosophy to that of "modern" laboratory science. Between 1667 and 1672 several of its members received royal employment or special *gratifications* as rewards for their intellectual endeavors.

Louis XIV's *approbation,* a talented membership, royal funding, and a coordinated research program—such attributes make the academy's failure a crucial chapter in the history of the "New Age of Academies." When the eighteenth-century scholar Bernard de Fontenelle attributed the start of that age to the "scientific renaissance of true philosophy," he launched a historiographic tradition that still shapes scholarship on French science.[4] Fontenelle saw a logical progression from the private-patronage organizations of the early 1660s to the founding of the Académie Royale; his modern adherents have developed that insight into more formal interpretations in which the Académie Royale stands as the natural culmination of almost a half century of organizational developments in French science.

[3]Traditionally, the academy in Montpellier (1706) has laid claim to the honor of becoming the "first" provincial adjunct to the Académie Royale. Obviously, in light of the Académie de Physique's existence, the Montpellier claim should be that it represented the "first successful" linkage of a provincial scientific academy to Paris. For discussion of the Montpellier academy, see Louis Dulieu, "La Contribution montpelliéraine aux recueils de l'Académie Royale des Sciences," *RHS* 11 (1958): 250–262, and "Le Mouvement scientifique montpelliéraine au XVIIIe siècle," *RHS* 11 (1958): 227–249.

[4]The best statement developing the full implications of the Fontenelle thesis is found in Hahn, *Anatomy of a Scientific Institution,* pp. 1–19. In effect, Hahn gave the Fontenelle thesis its formal definition as he crystallized a half century of historical research. Earlier works crucial to Hahn's argument include G. Bigourdan, *Les Premières Sociétés savants de Paris et les origines de l'Académie des Sciences* (Paris, 1919); Brown, *Scientific Organizations;* Pierre Gauja, "L'Académie Royale des Sciences (1666–1793)," *RHS* 1 (1949): 293–316, and "Les Origines de l'Académie Royale des Sciences de Paris," in *Académie des Sciences, institut de France, troisième centennaire, 1666–1967* (Paris, 1967), pp. 1–57; Albert Joseph George, "The Genesis of the Academy of Sciences," *AS* 3 (1938): 372–401; John Milton Hirschfield, *The Académie Royale des Sciences (1666–1683): Inauguration and Initial Problems of Method* (New York, 1981 [1957]); and René Taton, *Les Origines de l'Académie Royale des Sciences* (Paris, 1965).

To this day the "Fontenelle thesis" dominates the literature on the social and institutional history of French science. Such a situation is curious, especially in light of the considerable debates that characterize recent literature on English science. Historians have long recognized the need to define the exact nature of institutional arrangements in English science as crucial to an understanding of the greater issues in seventeenth-century science; yet these same historians hand down Fontenelle's basic premises on French science without subjecting them to critical examination.

The Académie de Physique's history allows us to begin just such a critical examination, for in it we find anomalies that are virtually impossible to reconcile within the broad interpretive generalities of the Fontenelle thesis. In fact, the Académie de Physique's history seriously challenges the explanatory power of the Fontenelle thesis. In order to understand why, we need to look first at the interpretive structure behind this historiographic tradition.

The strength of the Fontenelle thesis lies in the links it forges between Fontenelle's "scientific renaissance of true philosophy" and his "New Age of Academies." Was there a direct causal chain inexorably pulling a maturing French scientific community toward more formal scientific organizations during the 1660s? If so, what was it? Modern proponents of the Fontenelle thesis claim to find such a causal chain. Those who address the question agree unanimously on two points: first, French private-patronage science was in crisis during the early 1660s, and second, Colbert resolved the crisis in 1666 when he created the Académie Royale des Sciences. The Fontenelle thesis maintains that the Académie Royale filled the organizational vacuum left when private patrons abandoned the scientific community in its hour of greatest need—just as expensive, instrument-based laboratories became necessary to science.

Advocates of the Fontenelle thesis offer several pieces of evidence to make their case, but as we look closely at their arguments, they all come to rest on just two crucial points. First, they claim, by mid-1665 the last meaningful private-patronage academy in Paris (the Thévenot) had closed.[5] No new groups appeared before the opening of the Académie Royale eighteen months later. Thus an organizational vacuum supposedly existed for at least eighteen months—many have claimed the organizational vacuum lasted much longer, up to three

[5]Works that develop this point include Brown's *Scientific Organizations,* p. 136; George's "Genesis," p. 375; and Hirschfield's *Académie Royale,* p. 12.

years.[6] Second, between 1663 and 1666 various individuals in the Parisian scientific community made concerted efforts to persuade Louis XIV and his minister Colbert to found a royal academy of sciences. Advocates of a new royal institution (particularly Samuel Sorbière and Adrien Auzout) cited the expenses of the "new science" as beyond the means of private patrons.[7] They publicly argued that only a king or rich prince could afford the cost of scientific research. Historians who endorse the Fontenelle thesis take those pleas at face value.

In the course of tracing out the Académie de Physique's failure, we will come across numerous issues that raise questions about the adequacy of the Fontenelle thesis and the evidence used to buttress it. Moreover, among the anomalies we find, two points bear directly on the major evidential foundations underpinning the thesis.

First, as we shall see in Chapter 2, the Académie de Physique initially defined itself as an empirical research society during 1665 and 1666. Moreover, it did so by borrowing its program from the private-patronage Thévenot in Paris. The Fontenelle thesis requires meaningful private patronage to have disappeared from Paris before mid-1665 (without the organizational vacuum of 1665–1666, Fontenelle's rationale for the founding of the Académie Royale loses all force). Yet, as we shall see, one of the Académie de Physique's members was in Paris and attending Thévenot's weekly meetings during the second half of 1665 and the first months of 1666. These two organizations maintained regular contact until the moment Colbert announced the monarchy's intention to establish a royal institution (in March 1666). Only then did the Thévenot close. In fact, no organizational vacuum crippled French science during 1665 and 1666—either in Paris or in the provinces.

Second, during the Académie de Physique's life as a royal organization, it received the benefit of royal patronage and largesse that such Parisian propagandists as Sorbière and Auzout had asked for. Indeed,

[6]Among the works on the founding of the Académie Royale (see n. 2 above), only those of Brown, Hirschfield, George, and Hahn recognize a significant role for the Thévenot Academy in the organizational history of French science. All other sources date the failure of private patronage to the closing of the Montmor Academy in 1663 and extend the "organizational vacuum" from that point to the end of 1666.

[7]The documents in question are known as the "Sorbière Discourse" ("Cinq Cents de Colbert," BN 485, ff. 441–444) and the "Auzout Letter," which is actually a preface dedicating Auzout's *Ephemérèride du comète de 1664* to Louis XIV. The "Sorbière Discourse" has been reprinted in Bigourdan's *Premières Sociétés savants.*

the Académie became a state-funded institution. Still, it failed to establish itself as a viable entity. The academicians in Caen mouthed the same eloquent pleas for royal support as those Sorbière and Auzout voiced in Paris. The monarchy responded to those pleas; yet royal funding made no significant difference to the academy in Caen. As we put that fact together with the knowledge that the Thévenot continued in operation a full year longer than any historian of French science has ever recognized, we have a basis for questioning the Fontenelle thesis. If the Parisian patronage vacuum did not exist, and if royal patronage and financing did not guarantee a viable scientific institution, what can the Fontenelle thesis really tell us?

Unquestionably Fontenelle and his modern adherents are correct in identifying the "scientific renaissance of true philosophy" in the attempts to establish new organizational forms and new scientific programs during the 1660s. This was an era of flux in both the intellectual content and the organizational structures governing French science. It is also true that during this period new groups quickly formed and then just as quickly dissipated. Indeed, the Académie de Physique's own history offers dramatic evidence on how rapidly the fortunes of a private-patronage society could shift. Nevertheless, granting such "factual" components to the Fontenelle thesis does not validate its larger interpretive conclusions.

Whereas the strength of the Fontenelle thesis lies in the causal links it tries to forge between failing patronage and the founding of the Académie Royale, its heart lies in its assertion of a logical continuity between private-patronage and state support. It interprets the 1660s as a period of rapid transition, during which the natural development of sophistication among scientific practitioners created first disorder (when private patrons failed to continue their financial support), then order (when Colbert opened the Académie Royale). The key constructs in the historical interpretation it offers, then, form a paradox. The Fontenelle thesis presents the 1660s both as the crisis point at which French science passed dramatically from one stage to the next and as a period of natural and continuous progress.

From Fontenelle's vantage point in the eighteenth century, such an explanation made perfect sense. Fontenelle was writing the in-house history of the Académie Royale, looking back on its origins from a vantage point less than seventy-five years removed from the founding. He had the advantage of having personally known some of the individuals involved, but he also suffered the disadvantage of a limited perspective on the event. Fontenelle was interested only in his academy's origins, and when he came to explain them, he described the institution as a logical and natural culmination of events in the 1660s. He saw the Académie Royale as bringing order from the chaos of the early 1660s. In that sense, the Académie Royale—his institution—was the logical and inevitable product of an age that had tamed French society.[8] Like a Whig historian of the nineteenth century, Fontenelle proudly looked back on his past as inexorably guiding events to their glorious conclusion in his contemporary present. In Fontenelle's work, the Académie Royale—state-controlled science—represented the obvious culmination of a European scientific revolution that had begun in the sixteenth century. Until now, modern scholarship has done little to confirm (or to challenge) such an obviously parochial and whiggish view of French science.

The evidence the Académie de Physique's history brings to bear on the institutional history of French science cannot totally destroy (or rewrite) the Fontenelle thesis. Nevertheless, the academy's story raises the first serious questions about how much explanatory power the Fontenelle thesis actually contains. State involvement with the academy did nothing to secure its scientific future. With this organization, we see private patronage very much alive and well—part of both Parisian and provincial science—during the early and mid-1660s. At the very least, then, the Académie de Physique's history exposes the fundamental paradox at the heart of the Fontenelle thesis, which portrays state-controlled, state-funded science as both the resolution to an acute crisis and the logical and natural culmination of fifty years of institutional change. The Académie de Physique's history offers no easy resolution to that paradox. For this academy, state support created new problems; it did nothing to resolve old ones.

The academy's history explodes the myth of a patronage crisis in French science during the early 1660s. In the process it raises serious questions about whether royal involvement in seventeenth-century French science solved any of the immediate problems that faced the French scientific community during the seventeenth century.

[8]One of Fontenelle's most dramatic statements on this issue appears in his "Eloge de l'abbé Gallois": "Colbert favorisait les lettres, porté non-seulement par son inclination naturelle, mais par une sage politique. Il savait que les sciences et les arts suffiraient seuls pour rendre un règne glorieux; qu'ils lui donnent l'empire de l'esprit et de l'industrie, également flatteur et utile; qu'ils attirent chez elle une multitude d'étrangers, qui l'enrichissent par leur curiosité, prennent ses inclinations, et s'attachent à ses intérêts" (*Oeuvres de Fontenelle,* ed. G.-B. Depping, vol. 1 (Geneva, 1968 [1818]), p. 105.

1
The Academy's Origins

The Académie de Physique was a unique product of its provincial world. Originally it had no formal membership, no long-range scientific program, no fixed schedule for its meetings. Its founding followed from the personal ties of friendship, loyalty, and obligation that bound men together in seventeenth-century Caen.[1] Created within traditional forms, the Académie de Physique began as a patronage circle, or *assemblée*. For such an organization to exist, all that was needed was a host or patron who occasionally invited interested friends to his house for a "rare and curious" demonstration or a dissection. Such sessions simply gave these men a chance to gather for an evening's discussion of natural philosophy.

In its original form, the Académie de Physique de Caen was an unpretentious organization. Yet accidents of time and place made this informal circle the basis of one of the most ambitious projects for organizing science in the seventeenth century. A curious mixture of old and new, the Académie de Physique passed through a series of dramatic changes in less than a decade: starting from traditional

[1]Brennan's "Culture and Dependencies" is a valuable source on the personal nature of these relationships. A second work that is particularly important as background to the issues raised in this chapter is Tolmer's *Huet.* See also Gabriel Vanel, *Une Grande Ville au XVIIe et XVIIIe siècles: La Vie publique à Caen,* 3 vols. (Caen, 1910). Pierre-Daniel Huet's *Origines de la ville de Caen et des lieux circonvoisins,* 2d ed. (Rouen, 1706), and his *Pet. Dan. Huetii, Episcopi Abrincensis, Commentarius de Rebus ad eum Pertinentibus* (Amsterdam, 1718; hereafter *Commentarius*) remain important sources on learned Caennais in the seventeenth century. For further bibliography and a later history of the city, see Jean-Claude Perrot, *Genèse d'une ville moderne: Caen au XVIIIe siècle,* 2 vols. (Paris, 1975).

natural philosophy, its scientific program became an extreme, almost radical form of empirical science. Founded on traditional patronage, it was among the first academies to attempt a formal, cooperative structure in the practice of science. Within five years of its casual, almost fortuitous beginning, the king's minister Colbert took the Académie de Physique under royal protection and made it France's first royal academy outside Paris.[2]

The history of the Académie de Physique can tell us a great deal about the diffusion of ideas, knowledge, and organizational forms in seventeenth-century France. Indeed, its history offers an extraordinary case study. Its founders were *curieux*, amateur natural philosophers eager to create a new form of empirical science. These men tried to make their *assemblée* a part of the New Age of Academies. Their motives reveal a great many things about that age. More specifically, their story tells us about the enthusiasm for new scientific organizations during the 1660s, while also revealing the frustrations the founders of any new scientific organization must have faced. More than anything, the history of the Académie de Physique is the story of a few men who tried to change their world.

THE FOUNDERS

Two men created the Académie de Physique: Pierre-Daniel Huet, one of the great French savants of the seventeenth century,[3] and

[2]Numerous private academies were created in seventeenth-century France (for example, Moisant de Brieux's Grand Cheval was created in Caen in 1652), but very few were granted a royal charter before the eighteenth century. With the notable exception of the Académie des Jeux Floraux in Toulouse, however, no provincial academy could claim any form of royal political legitimation for its existence before the Académie de Physique. Even the Jeux Floraux, which claimed royal privileges dating back to 1323, was given its "modern" royal charter only in 1694. The first provincial academy to receive letters patentes was the academy in Soissons (1674). Those letters patentes are published in *LIMC*, 5:550–551. For a chronology of the founding of provincial academies, see Daniel Roche, "Milieux académiques provinciaux et sociétés lumières," in *Livre et société dans la France du XVIIIe siècle*, ed. F. Furet (Paris, 1965), pp. 93–184. For an extensive discussion of the eighteenth-century academies, see Daniel Roche, *Le Siècle des lumières en province: Académies et académiciens provinciaux, 1680–1789*, 2 vols. (Paris, 1978).

[3]Pierre-Daniel Huet (1630–1721) has been the subject of numerous biographies and specialized studies. The best general work on his life is Tolmer's *Huet*. For more recent bibliographies dealing with specialized aspects of Huet's life, see Brennan, "Culture and Dependencies," and Richard Popkin's article "Pierre-Daniel Huet," in *Encyclopedia of Philosophy*, ed. Paul Edwards, 8 vols. (New York, 1967), 4:67–68.

André Graindorge, a provincial *médecin* who left the world little beyond a few minor scientific writings.[4] These were two very different men. Once circumstances brought them together, however, their relationship could, and did, provide an almost ideal basis for their scientific organization.

Huet and Graindorge lived in a world where *ordres, états,* and *fidélité* organized social relationships.[5] Theirs was a hierarchical society. They shared common interests in natural philosophy, but they were not equals.[6] When they began their association in 1660, Pierre-Daniel Huet was an important figure in a circle of learned Caennais whose collective erudition prompted contemporaries to describe their city as the "modern Athens." Just thirty years old in 1660, Huet had been counted among Caen's learned elite for a decade and had already earned a reputation as a poet and philologist.[7] Moreover, he had begun his project for producing a critical edition of Origen's commentary on Matthew, a work that would soon mark him as one of the leading savants in the French Republic of Letters.[8] Young, noble, and

[4]Graindorge's life and scientific writing are discussed in Tolmer, *Huet.* See also Brown, "L'Académie de Physique de Caen."

[5]For specific discussion of such relationships in seventeenth-century France, see Roland Mousnier, "Les Concepts d' 'ordres,' d' 'états,' de 'fidélité' et de 'monarchie absolue' en France de la fin du XVe siècle à la fin du XVIIIe," *Revue Historique* 247 (1972): 289–313; as well as Pierre Lefebvre, "Aspects de la 'fidélité' en France au XVIIe siècle: Le Cas des agents des princes de Condé," *Revue Historique* 250 (1975): 59–106. For more general discussions of stratification theory and its relevance to early-modern Europe, see Gerhard E. Lenski, *Power and Privilege: A Theory of Social Stratification* (Chapel Hill, N.C., 1966); Roland Mousnier, *Les Hiérarchies sociales de 1450 à nos jours* (Paris, 1969); and William H. Sewell, "Etat, Corps, and Ordre: Some Notes on the Social Vocabulary of the French Old Regime," in *Socialgeschichte heute: Festschrift für Hans Rosenberg zum 70. Geburtstag* (Göttingen, 1974), pp. 49–68.

[6]Both sprang from families of Protestant bourgeois that had risen to prominence in Caen during the sixteenth century, but within their world Huet was superior to Graindorge in every meaningful sense. Huet claimed illustrious parents and noble birth; Graindorge sprang from a family of artisan weavers and achieved nobility (sieur de la Londe) only in 1653. Huet was born in the Catholic faith; Graindorge converted as an adult. (See Brown, *Commentarius,* p. 3; "L'Académie de Physique," pp. 125–126.) Huet was an eldest son and thus the family heir, Graindorge a cadet with two heirs standing between him and the family patrimony. Finally, at the time they met Huet had already acquired considerable *état* within the Republic of Letters; Graindorge was almost unknown.

[7]The best source on Huet's intellectual activities during the 1650s is his *Commentarius,* pp. 71–218. See also Tolmer, *Huet,* pp. 125–267.

[8]Huet's *Origenis in Sacras Scripturas Commentaria, Quaecunque Graece reperiri Potuerunt,* 2 vols. (Rouen, 1668). For discussion of this work, see Tolmer, *Huet,* passim, and Brennan, "Culture and Dependencies," pp. 63–64.

master of his own fortune, in 1660 Pierre-Daniel Huet was already a man of accomplishments.[9]

André Graindorge was already forty-four years old in 1660, and although a longtime dabbler in natural philosophy, he as yet had no real scholarly or intellectual accomplishments to his name. By any reckoning he held a lowly position in the social and intellectual hierarchies of the day. Though he had been born in Caen, Graindorge was virtually unknown there. The second son in a Protestant family, he had been sent to the university at Montpellier, probably about the time Huet was born, and then for more than twenty years had practiced medicine in Narbonne. Only his family responsibilities had brought him back to Caen in 1660. His older brother had died, leaving two minor children, and Graindorge had been called back by their *tutelle*. Many contemporaries considered such responsibility burdensome, but for an unpropertied provincial physician such as Graindorge it presented an opportunity. Although he was not the heir to his brother's estate, the guardianship over his nephews made him the provisional head of his family, giving him control over the family lands and fortune during the boys' minority. In that way he gained the time and resources for his studies in natural philosophy.[10]

To Graindorge, Huet was *un grand personnage*—someone whose *état* brought royal pensions, access to the king and his ministers, or admission to Parisian academies.[11] Graindorge could not hope to acquire such greatness on his own. He could aspire to local prominence (he did in fact become an *échevin de ville*),[12] but Graindorge

[9]The best gauge of contemporary opinion concerning Huet's "promise" is found in the records of the "Gratifications faites par Louis XIV aux savants et hommes de lettres français et étrangers, de l'année 1664 à l'année 1683" (*LIMC*, 5:466–497). With the exception of the years 1673–1677 (during which Huet was in royal service both as *sous-précepteur au dauphin* and as a member of the Académie Française), he regularly received a pension of 1,500 livres a year. For Chapelain's assessment of Huet's talents in the early 1660s, see his "Liste de quelques gens de lettres François viviants 1662," BN, Fr 23045, ff. 107r, 116v.

[10]For discussion of Graindorge's life before 1660, see Brown, "L'Académie de Physique," pp. 125–127. For a general discussion of the social status of physicians, see François Millepierres, *La Vie quotidienne des médecins au temps de Molière* (Paris, 1964).

[11]Huet received his first royal pension in 1663, his first audience with Louis XIV in 1667 or 1668, and admission to the Académie Française in 1674 (Chapelain to Huet, 18 August 1663, BL, A 1866, 244; *Commentarius*, pp. 245, 308–309).

[12]Brown, "L'Académie de Physique," p. 125. For discussion of the city government of Caen, see Jean Yver, "La Ville de Caen: Le gouverneur et les premiers intendants de 1636 à 1679," *MANC*, n.s. 7 (1934): 313–371.

realized that he had limitations. He knew that his only hope of real success as a *curieux* lay in attaching himself to a great and powerful man within the Republic of Letters—someone like Huet. That is exactly what he did. Huet's taste for natural philosophy presented Graindorge with an opportunity. With this interest as their common ground, André Graindorge quickly became Huet's "faithful," or *fidèle*. In return, Pierre-Daniel Huet became Graindorge's patron and protector within the Republic of Letters.

The origins of the Académie de Physique lay in this patron–client relationship. By accepting Graindorge's loyalty as a *fidèle*, Huet committed himself to furthering Graindorge's ambitions.[13] Huet fulfilled his obligation during the 1660s, and as a result, these two men created the Académie de Physique. In short, André Graindorge's ambition to become a scientist was the driving force behind scientific activities at Huet's house. Huet's patronage furnished the mechanism that allowed him to fulfill his ambition.

The patron–client relationship between Huet and Graindorge explains how these men established the Académie de Physique. Nevertheless, these men were not stock actors playing out the roles of patron and *fidèle*. Nor did the academy emerge in lockstep from their friendship. Each man brought his own capabilities and limitations to the relationship. They had very different reasons for becoming involved with natural philosophy, and very different personalities. Such individual differences played a crucial part in shaping the academy and must be accounted for. Theirs was a dynamic relationship, and in order to understand how and why it worked, we must first understand each of these men individually.

Pierre-Daniel Huet, Patron

When he met André Graindorge in 1660, Pierre-Daniel Huet had already organized his life around a task that was going to dominate his attention throughout his involvement with the Académie de Physique. Since 1652 he had been working to produce a critical edition of Origen's commentary on Matthew. As originally conceived, this project involved a relatively simple process of transcription, translation, and editing, requiring only a few months to complete.[14] Yet in 1660

[13]For discussion of the force behind such a commitment, see Sharon Kettering, *Patrons, Brokers, and Clients in Seventeenth-Century France* (New York, 1986).

[14]*Commentarius*, p. 107. For Huet's own statement on his commitment to the project, see ibid., p. 151.

Huet was still in the first stages of the project. In fact, he was still working on a preliminary treatise explaining his approach to the text.[15] Huet's slow progress was not caused by laziness or a lack of effort; on the contrary, he devoted extraordinary energy to it. By the early 1660s the Origen project was consuming more and more time with each passing year because it had become controversial. Many contemporaries saw Huet's simple, straightforward critical edition of a patristic text as a sectarian religious tract.[16]

The Origen project conditioned all of Huet's involvements with natural philosophy, Graindorge, and the Académie de Physique. Huet was an ambitious man.[17] As he tells us in his autobiography, the *Commentarius,* he had a passion for belles lettres from earliest childhood. Orphaned young, he was in his late teens by the time he decided to ignore his guardian's wishes that he become a lawyer and *parlementaire,* and determined instead to make his career in the Republic of Letters. In 1650, when he reached his majority and received his inheritance, he set about doing just that.[18] Huet's ambitions were directed toward Paris. He wanted acceptance within the highest circles of the Republic of Letters. Yet after some initial forays in Paris during the early 1650s, Huet settled down to spend the remainder of the decade and most of the early 1660s in provincial Caen.[19] In 1666 he resumed the pursuit of his career in Paris, spending more time there than in his native city over the next four years. In 1670, when he became *sous-précepteur* to the Dauphin, he closed his house in Caen and left the city altogether.[20]

What kept Huet in Caen throughout the late 1650s and early 1660s was the Origen project. At the time Huet received his inheritance (January 1650), he intended to launch his career in Paris and then to broaden his intellectual horizons by travel.[21] Between 1650 and 1652 he obtained introductions to the learned circles in Paris, and in early

[15]*De Interpretatione Libri duo: Quorum prior est de optimo Genere Interpretandi: Alter de Claris Interpretibus* (Paris, 1661).

[16]*Commentarius,* pp. 150–151.

[17]For an excellent discussion tracking Huet's ambitions through the 1660s, see Brennan, "Culture and Dependencies," pp. 62–76. For a less charitable portrayal of Huet's careerism, see Vanel, *Une Grande Ville,* 2:316–322.

[18]*Commentarius,* pp. 15–44, 57–40.

[19]Despite his claim of frequent visits to Paris (*Commentarius,* p. 166), Huet's record of daily expenses (BN, Fonds Français, nouvelle acquisitions 1197) shows he could not have been in Paris between April 1655 and June 1659.

[20]For a detailed discussion, see Tolmer, *Huet,* pp. 400–403, 417–424.

[21]*Commentarius,* pp. 58–59, 72.

1652 he was planning to visit Italy when instead, at the last minute, he joined the entourage of promising young scholars who accompanied Samuel Bochart on his journey to the court of Christina of Sweden.[22] That last-minute decision proved a major turning point in his life.

Things did not go well for the Bochart party in Sweden. Christina had invited Bochart to her court, but between the time she sent her invitation and Bochart's arrival a year later Christina had undergone a change of heart. Under a variety of pressures, including poor health, her plans for conversion to Catholicism, and the opposition of the Swedish nobility, who were becoming increasingly resentful of royal largesse "wasted" on foreigners, Christina had curtailed her scholarly activity. [23] When the Bochart party arrived, Christina did little to indicate she was the "Minerva of the North" who had lavished her attention so recently on René Descartes, Claude Saumaise, and Isaac Vossius. In fact, Bochart had difficulty obtaining interviews with the queen.[24]

For Huet—a Catholic traveling with Calvinists to a Lutheran court—the lack of royal favor toward the Bochart party made the situation particularly uncomfortable. When Huet discovered a copy of Origen's commentary on Matthew among Christina's manuscripts, Bochart suggested that his young protégé produce an edition. Christina consented and wanted Huet to complete the project in Sweden under her protection. According to Huet's *Commentarius*, he agreed to Christina's conditions reluctantly, asking only that she allow him to return to Caen for a short time to attend to some personal affairs. His autobiography then claims that the prospect of Christina's abdication prevented his return.[25] The contemporary evidence, however, offers only limited support for his account. Apparently Huet made an unauthorized copy of the Origen manuscript and then, abandoning the Bochart party, left Sweden with the manuscript but without the

[22]For Huet's accounts of his reasons for joining the Bochart party, the problems in Sweden, and the effects on his own life, see *Commentarius*, pp. 71–139. For discussion of Bochart's life and his place in the intellectual community of Caen, see Edouard-Herbert Smith, "Recherches sur la vie et les principaux ouvrages de Samuel Bochart," in *MANC* (1836), pp. 341–377. For a summary and brief description of Bochart's ideas, see Katherine Brownell Collier, *Cosmogonies of Our Fathers* (New York, 1968 [1934]), pp. 63–67.

[23]For a description of Christina's court, her political problems, and her patronage, see Georgina Masson, *Queen Christina* (New York, 1968).

[24]*Commentarius*, pp. 103–105.

[25]Ibid., pp. 105–115.

knowledge or consent of Christina.[26] Nothing suggests that Huet intended to return.[27]

Once back in France, Huet made a significant change in his approach to a career. Instead of seeking his fortune in Paris, or in Sweden or Italy, Huet settled down to spend most of the next fifteen years in Caen. He never did visit Italy, and he claimed he refused Christina's offer of "mountains of gold" if he would join her entourage.[28] He did not even venture as far as Paris again until 1659.[29] Even after he had secured the favor of a powerful patron in Paris (Jean Chapelain), he stayed in Caen for another decade.[30] Huet's new attachment to provincial Caen can be explained in part by the decision to make the Origen edition his chef d'oeuvre, but it seems that this decision was itself affected by another change in his life. While Huet was in Sweden, one of the leading figures in the circles of learned Caennais, Jacques Moisant de Brieux, had established an academy of belles lettres.[31] This organization, which was called L'Académie du Grand Cheval, gave new structure to the savants in Caen.[32]

Almost as soon as Huet arrived back in Caen, Moisant de Brieux paid him a call and told him that he had been elected to the new academy.[33] For someone as young as Huet, such attention from Brieux was truly an honor. Moreover, for someone with his ambitions, election to the Grand Cheval offered a genuine opportunity. At the time (early 1653), he was only twenty-three years old and without

[26]Bochart to Huet, 2 November 1652, BL, A 1866, 61. In this letter Bochart told Huet that he had informed the queen of Huet's departure. His account of the queen's reaction to that news indicates that Christina had no idea Huet was planning to leave Sweden.

[27]In fact, Huet admits as much in *Commentarius*, p. 113: "Quamobrem etsi fidem meam de reditu obligaveram Reginae, nulla mihi tamen religio fuit, Holmia proficiscenti, cum votum Mercurio pro reditu in patriam nuncuparem, nunquam me in Sueciam regressurum."

[28]*Commentarius*, p. 208.

[29]See n. 19 above.

[30]For Huet's own account of his relationship with Chapelain, see *Commentarius*, pp. 160–161. See also Tolmer, *Huet*, pp. 231–233. For biographical information on Chapelain, see G. Collas, *Jean Chapelain* (Paris, 1911). For discussion of his interests in science, see Albert Joseph George, "A Seventeenth-Century Amateur of Science, Jean Chapelain," *AS* 3 (1938): 217–236.

[31]For biographical information on Brieux, see René Delorme, "Moisant de Brieux, fondateur de l'académie de Caen," in *MANC* (1872), pp. 27–110.

[32]Brennan's "Culture and Dependencies" discusses Brieux and his academy.

[33]*Commentarius*, p. 142.

either substantial scholarly accomplishments or a powerful patron. In that situation, the Grand Cheval offered Huet more than he could hope to find in Paris. The only formal organization in Paris comparable to Brieux's Grand Cheval was Seguier's Académie Française, and it would be years before Huet could hope to qualify for election to that body.[34] Indeed, by entering the Grand Cheval, Huet began to build the scholarly career that led him to the Académie Française twenty years later. In 1653 Huet could establish his career without going to Paris; Moisant de Brieux's new academy had brought the Republic of Letters to Caen.

Huet termed his election to the Grand Cheval one of the "great honors" of his life.[35] The event certainly had a dramatic impact on his life: the Origen project took form there. Reflecting the heterogeneous intellectual community in Caen, the Grand Cheval contained a healthy contingent of Protestants, including Brieux himself. In such an atmosphere, the young Catholic Huet had to defend his handling of the Origen manuscript at every step of the way. Slowly the project expanded from the simple translation conceived in Sweden to become a major theological work. In fact, almost as soon as he started, Huet put the Origen manuscript itself aside and began a philological treatise justifying his interpretive methods.[36] Altogether, from its conception to the publication of the final work as *Origenis Commentaria* (1668), the project consumed sixteen years. And as the project grew, Huet found that he sometimes needed relief from the frustrations of an intellectual task that took years, not months, to complete. At such times he turned to natural philosophy.[37] After 1660 he turned to his *fidèle*, Graindorge.

Huet faced genuine frustrations in the Origen project, and as late as 1665 he considered abandoning it. Indeed, during March of that year he committed his reservations to paper, claiming both that he considered the Origen "bas & obscur" and that he found greater

[34]In 1674 Huet was elected to the chair originally held by Marin Le Roy de Gomberville.

[35]*Commentarius*, p. 142.

[36]*De Interpretatione* (1661).

[37]For discussion of how Huet moved back and forth between natural philosophy and the Origen project during the late 1650s and early 1660s, see Tolmer, *Huet*, pp. 177–228. Tolmer's descriptions of these oscillations between two forms of intellectual activity are excellent, but his general conclusion that this is simply evidence of a broad-ranging and voracious intellect cannot be accepted.

satisfaction in his newfound work with astronomy.[38] Yet the *Origenis Commentaria* was the chef d'oeuvre that established him as a "master" within the Republic of Letters, even though from the start it alienated Huet from erstwhile friends and supporters. In fact, Huet's handling of the Origen manuscript very quickly caused an irreconcilable break with Samuel Bochart, his first real patron in the Republic of Letters and the man who had suggested the project.[39]

The Origen project, then, led Huet to an intellectual crisis, which caused him to turn more seriously to natural philosophy.[40] Huet's newfound interest surprised and pleased his *fidèle*. It occurred at a fortuitous moment in Graindorge's own career—just as he was coming to believe that scientific academies would bring civilization to a new Golden Age of Science, what he called a *siècle-d'or*. Graindorge believed that Huet's active participation in the Académie de Physique would ensure that organization a glorious future. Before dealing with Graindorge's grand plans, however, we must first understand this man who saw his patron as the key to a successful scientific organization.

André Graindorge, *Fidèle*

André Graindorge was a man of the seventeenth century. His words, like those of a Molière character, can shock and surprise the modern reader. Sometimes pompous, sometimes fawning, he more often delights us with pithy insights into the universal feelings behind human actions. But perhaps to compare Graindorge with one of Molière's comedic inventions is to do him a disservice. Neither a buffoon nor a fool, Graindorge, like his contemporary Molière, possessed a highly developed sense of the social conventions governing his world. He knew that whatever he wanted to accomplish, he must

[38]Huet to Chapelain, 5 March 1665, BL, A 1866, 2408, 2409, and 2505. These three *inserti* represent successive drafts of the same communication.

[39]For Huet's account of this break, see *Commentarius*, pp. 149–150.

[40]A succinct summary of Huet's philosophical development can be found in Popkin, "Pierre-Daniel Huet," 4:67–68. More detailed accounts can be found in J. d'Avenal, *Histoire de la vie et des ouvrages de Daniel Huet, évêque d'Avranches, ou Le Scepticisme théologique* (Paris, 1850); A. Durpront, *P.-D. Huet et l'exégèse compariste au XVIIe siècle* (Paris, 1930); Richard H. Popkin, "The High Road to Pyrrhonism," *American Philosophical Quarterly* 2 (1965): 18–32, and "The Sceptical Crisis and the Rise of Modern Philosophy," *Review of Metaphysics* 7 (1953–54): 132–151, 307–322, 499–510.

operate within the social forms laid down for him. For Graindorge, those forms mandated a patron. He wanted a "career" as a natural philosopher/scientist. Without a patron, he felt powerless. As Huet's *fidèle,* he waxed confident, even aggressive.

Graindorge's strategy for pursuing his career as a natural philosopher was rooted in a recognition of how the Republic of Letters worked. Confident in his own talents, knowledge, and abilities, he knew he was at least Huet's equal as a scientist. He corrected Huet's "false" opinions and sometimes used condescendingly didactic terms to instruct his patron in laboratory procedure. Indeed, in the laboratory Graindorge became Huet's mentor. Yet Graindorge consistently saw his fate as hanging by the slender thread of patronage. In one of his first letters to Huet, for example, Graindorge thanked his patron for dedicating a scholarly publication to him (in memory of Jacques Graindorge de Prémont), saying: "I could hardly have believed the stature you have fixed on me. . . . I shall blush . . . [but] I shall raise my eyebrows and carry my head so high that I shall be taken for a great man [*je passerai pour un grand personnage*] . . . you will not regret having elevated me so high when I put on my face as a philosopher."[41] This passage reveals a Graindorge who knew he was no "great man," but also one who trusted his abilities as a scientist to sustain his relationship with Huet, who was indeed *un grand personnage.*

Given the social realities of seventeenth-century France, Graindorge's attitudes toward Huet, patronage, and his own position made sense. From top to bottom, this society of "orders" organized itself around the principles of personal subordination, service, and loyalty. Within the Republic of Letters, these principles powered the patronage system. With his personal revenues, for example, Pierre-Daniel Huet did not "need" Chapelain's patronage in a direct economic sense, nor did he need Chapelain for advancement in any broad social sense. Nevertheless, Huet sought Chapelain's favor as carefully as if his subsistence and title depended on it.[42] Chapelain's favor marked Huet with the *état* he needed in the Republic of Letters. Moreover, as Chapelain's *fidèle,* Huet stood just two steps removed from the king himself. Success required patronage, and thus Graindorge's approach to his career resembled Huet's. After all, as Huet's

[41]VDLI, pp. 261–266 (24 April 1661).

[42]For an extensive discussion of the kinds of services Chapelain rendered various Caennais, see Brennan, "Culture and Dependencies," pp. 86–134.

fidèle, Graindorge stood on the same patronage ladder–only three rungs removed from Louis XIV.[43]

Like Huet, Graindorge was a member of the Grand Cheval. Graindorge, however, did not share Huet's commitment to Brieux's academy. Willing to admit (grudgingly, perhaps) that his world valued belles lettres over natural philosophy and his own empirical science, Graindorge nonetheless had difficulty finding much to his taste at the Grand Cheval. In one of his first letters to Huet, for example, he waxed enthusiastic about a scientific problem Huet had found in Paris, then offhandedly summarized the discussion at the most recent session of the Grand Cheval for his patron: "They talked at our last session about I don't know what lotion used by women in former times. . . . They presented diverse passages, going all the way back to the Hebrew."[44]

Graindorge had little patience for what he considered the dry, arcane discussions at the Grand Cheval; nor, as a convert to Catholicism, did he approve of what he perceived to be the heavily Protestant bias at Moisant de Brieux's academy.[45] Although he attended sessions and participated in philological efforts to trace social customs to their ancient sources, Graindorge longed to find "modern authors working on philosophy, not those interested in pedantry, but those working to discover something new."[46]

Graindorge probably inherited his chair in the Grand Cheval from his brother Jacques Graindorge de Prémont.[47] If so, it fitted him poorly. As with the *tutelle* of his nephews, the redeeming attraction of the Grand Cheval was that it involved him with Huet. In fact, he attended as Huet's *fidèle.* When Huet traveled, Graindorge's letters kept him informed, and he often relayed greetings between Brieux and Huet. When asked, he voiced Huet's opinions or read treatises Huet had sent for presentation. He willingly inconvenienced himself to carry out these missions for his patron—as when he agreed to

[43]For discussion of Chapelain's relationship with both Huet and Colbert, see Collas, *Jean Chapelain.* For discussion of Colbert's relationship with Louis XIV, see Inès Murat, *Colbert* (Paris, 1980), pp. 116–128. For an excellent general discussion of Louis XIV's impact on "patronage," see John B. Wolf, *Louis XIV* (New York, 1968), pp. 133–136.

[44]VDLI, pp. 255–261 (11 April 1661).

[45]For example: "L'on ne parla qu'hébreu à notre dernière assemblée. Un certain Morin [a Protestant minister] s'y trouva, qui y fut accueilli avec acclamation par les frères. Il débita de sa marchandise" (VDLI, pp. 261–266 [24 April 1661]).

[46]VDLI, pp. 255–261 (11 April 1661).

[47]Prémont had been one of the original five who helped found the academy. See Brennan, "Culture and Dependencies," pp. 38–39; *Commentarius,* pp. 52–53.

search out a particular epitaph: "I judge it deserves the effort since you intend it for such a clever group. Nothing we have seen so far [from Brieux's academicians] is worthy of the Pont-Neuf."[48] Simultaneously serving and flattering, the insult in rating Brieux's academicians below the scurrilous poets of the Pont-Neuf could not have been lost on Huet. Graindorge never disguised his contempt for the Grand Cheval from Huet. He attended, but only to secure Huet's patronage.

Graindorge wanted to practice a very particular form of natural philosophy, one that deemphasized traditional philosophical systems and relied instead on observation and experiment. In effect, he wanted to practice the "new science," and like other seventeenth-century advocates of the "new science," he faced a dilemma. Most educated people still considered strictly empirical, laboratory science an activity beneath the dignity of great and learned men; yet Graindorge thought savants such as Huet the men best equipped to hasten the coming of a new Golden Age–his *siècle-d'or.* On the one hand, Graindorge believed that this new form of natural philosophy promised a glorious future of exciting discoveries that would sweep away past errors. He knew it was possible to produce new knowledge. On the other hand, he also believed that this *siècle-d'or* must await the establishment of his science as a respectable activity. Caught between his vision of the future and his respect for the great men of his age, he wanted these men to abandon their own *siècle des moeurs* to create his *siècle-d'or.*[49]

Graindorge was a reformer. He saw that in his world the success of the scientific revolution depended on the establishment of his science within the Republic of Letters. It was toward that end that he solicited Huet's support (and, later, royal patronage) in provincial Caen, just as his contemporaries Samuel Sorbière and Adrien Auzout tried to elevate the dignity of science in Paris by seeking Louis XIV's patronage. Graindorge, like Sorbière and Auzout, sometimes argued science's material needs in seeking patronage,[50] but in Graindorge's

[48]VDLI, pp. 155–261 (11 April 1661).

[49]Graindorge's most direct statement on this subject appears in his letter to Huet of 16 December 1665 (BL, A 1866, 547). In this letter he bemoaned the fact that "lart et long et la vie courte" and claimed his own age had not attained "la moindre verité" from the "siècle des moeurs."

[50]Graindorge's position on this question varied according to time and circumstance. As we shall see in chap. 4, he and the academicians in Caen publicly claimed they needed substantial funding to function as a royal academy. Nevertheless, in his private exchanges with Huet, Graindorge consistently minimized the academy's expenses,

case at least, that argument did not always reflect the actual state of affairs at the academy, nor did it ever reflect personal greed. Over the course of his involvement with the Académie de Physique, for example, there is no evidence that he profited, or hoped to profit, from his position as the academy's secretary. On the contrary, his scientific activities drained his personal resources. Indeed, in 1672 he misrepresented the academy's expenses, paying costs from his own pocket, to perpetuate the illusion that Pierre-Daniel Huet was continuing to support the academy.

André Graindorge thus sought to elevate himself by elevating his chosen milieu. A complex and intriguing person—very much a man of the seventeenth century—he accepted the social order that governed his life and had great respect for the traditional forms and organization of knowledge. Nonetheless, he was a man who could develop a utopian vision of the future offered by a new form of intellectual activity. Eager to hasten the coming of his *siècle-d'or* yet aware that his science was held in low esteem by most of his contemporaries, he sought to raise science's status or *état*. Huet's patronage allowed him to do so. Besides the luster provided by Huet himself, Huet's *état* brought other men of "quality" to the Académie de Physique, men Graindorge could not have recruited on his own. An effective strategy as long as he could count on Huet's involvement, Graindorge's reliance on patronage meant that the academy would have difficulty surviving Huet's withdrawal from active participation. When Huet did withdraw, Graindorge had to find other ways to hold the Académie de Physique together. His attempts to do so are what make the academy's story worth telling.

The history of the Académie de Physique is indeed the story of Graindorge's efforts to find patronage for his vision of a new form of natural philosophy. In that sense, it is a history that is exceptionally well documented. We learn a great deal about Graindorge, his ideas, and his efforts to secure patronage from the letters he addressed to Huet during the course of their association. Unfortunately, however, the Graindorge letters do not give any direct information on the question we must now address: How was the academy founded? On

indicating that aside from providing candles and firewood for the academy's sessions, Huet incurred no expenses as patron. Graindorge's consistent omission of exact figures for the cost of various items makes it extremely difficult to say anything more than that in his exchanges with his patron, the academy's finances were never discussed as a problem.

that issue Pierre-Daniel Huet's writings provide the most useful information.

FOUNDING THE *ASSEMBLÉE*

Pierre-Daniel Huet published two brief accounts of the history of the Académie de Physique. The first appeared in his *Origines de la ville de Caen et les lieux circonvoisins* (1702), the second in his autobiography, the *Commentarius* (1718). Curiously, these accounts appear to disagree on every question we might want to ask about the academy's origins: When and how did the founding occur? Who was involved? And why was the academy created? In fact, Huet's accounts offer two entirely different versions of the academy's early history; yet, given Huet's role as the academy's founding patron, all attempts to discredit his authority in either version have proved futile.[51] That being the case, we shall not try that approach here. Instead, we shall accept every statement Huet made in both accounts, then reconstruct the academy's early history by qualifying his interpretation of events. The key to this approach lies in recognizing the different purposes for which he wrote his two versions of the Académie de Physique history.

Despite Huet's common authorship, the *Commentarius* and *Origines* are very different works and represent distinct categories of historical literature. As an autobiography, the *Commentarius* is filled with self-justifications and rambling personal reminiscences. The *Origines de la ville de Caen* is a formal history of the city and its institutions. Neither work offers a straightforward chronicle of the academy's history, and strictly speaking, neither is a primary source. To use them thus one must discriminate between the testimony of Huet the historian and the more subjective account of the memoirist. Once we make that

[51]The problem of reconciling these two works has been raised (at least implicitly) in a number of works: L. G. Pellissier, "À travers les papiers de Huet," *Bulletin du Bibliophile*, 1889, pp. 25–27; Vanel, *Une Grande Ville*, 2:316–322; Brown, *Scientific Organizations*, pp. 216–220, and "L'Académie de Physique," pp. 127–131; Tolmer, *Huet*, p. 274, and VDLI, pp. 248–249; Brennan, "Culture and Dependencies," pp. 141–142. Pellissier, Brown, and Tolmer accept the account given in Huet's *Commentarius* and thus date the academy's origins to 1662 and give dissection as its purpose. For this group, particularly Tolmer, a letter Huet addressed to a nephew in 1702 offers decisive evidence. In that letter (Bibliothèque de Caen, ms. in-4, vol. ii, f. 241), Huet claimed he had founded the academy "vers l'année 1662." Vanel and Brennan accept the version given in *Origines de la ville de Caen* and thus date the academy to 1664, citing astronomy as its purpose.

discrimination, Huet's accounts reveal essential information about the academy's history.

In the *Commentarius* Huet tells his readers that he and Graindorge acted alone in establishing the Académie de Physique in 1662.[52] According to this version of events, Huet and Graindorge first agreed on a scientific agenda and a schedule for the meetings, then invited others to participate. In describing the early activities, Huet emphasizes a program of dissection and comparative anatomy. To explain the need for this new organization, he complains that the academicians of the Grand Cheval had "yawned" and acted "rude" whenever he had made presentations on natural philosophy there. Equally important, in the *Commentarius* version of events Huet stresses André Graindorge's role in founding the Académie de Physique. He claims that Graindorge was an experienced natural philosopher, credits Graindorge with suggesting the idea of a separate academy, and tells the reader that Graindorge took charge of organizing the sessions. In sum, in his autobiography Huet places the founding in 1662, describes Graindorge as the moving force behind the academy, and portrays himself as a patron endorsing the project of his *fidèle*.

The *Origines* tells an entirely different story.[53] In this history of Caen, Huet claims that a small group from within Moisant de Brieux's Grand Cheval first began to observe the comet of 1664–1665, then formed an academy, and finally "chose" him to host their scientific organization. In striking contrast to the *Commentarius,* Huet's *Origines* makes no mention of Graindorge. With regard to the scientific program, the *Origines* presents very little detail about the academy's activities but makes it clear that the academy was a full-fledged scientific organization from the start and claims that its members pursued research in several areas. Finally, in the *Origines* Huet places heavy emphasis on the academy's organization and its later problems with royal incorporation. In summary, then, the *Origines de la ville de Caen* puts the founding in 1664–1665 and describes the academy as a formalization of group interests. In this history of Caen, Huet is no longer the patron described in the autobiography; instead, he has become the elected host of a new scientific organization.

The fundamental difference between these two versions of the academy's history lies in Huet's selective use of evidence. For virtually every claim he made about the academy in either his autobiography

[52]The following account is based on *Commentarius,* pp. 219–220.

[53]The account presented here is based on *Origines de la ville de Caen,* p. 173.

or the *Origines* one can find some corroboration in contemporary documents.[54] For instance, in late 1670 Huet told Henry Oldenburg that the Académie de Physique had been meeting at his house for "eight years," a statement that clearly supports the *Commentarius* and its claim for a 1662 founding.[55] But André Graindorge's letters from the mid-1660s demonstrate that formally organized scientific activity at Huet's house did not begin until 1665, thus supporting the *Origines*. In short, Huet's two versions of the academy's history disagree on how and why it was founded, but each version is obviously based on events that occurred.

A reconciliation of Huet's accounts requires us to give due weight to every verifiable claim he made—in both accounts. When we do so, the broad outlines of the academy's early history begin to emerge. These accounts actually describe two stages in a process of organizational development. The *Commentarius* treats what we might call the "origins" of the Académie de Physique, and the *Origines de la ville de Caen* deals with its "regularization," or reorganization, as a private scientific society.

Whenever Huet's discrepancies appear to make the two accounts irreconcilable, a careful reading of what he actually says renders their differences comprehensible. For example, the disagreement over a 1662 versus a 1664–1665 founding unravels on close examination. Along with the claim for a 1662 founding in the *Commentarius,* Huet also asserts that this new academy, "which sprang from meager beginnings, continued to grow and later flourished among the more splendid."[56] In the *Commentarius* itself, that statement merely offers the vague hint of organizational development after 1662; placed against the claim for a later founding in the *Origines,* however, we can understand this same statement as revealing the extremely informal nature of the early *assemblée*.

Ironically, Huet's insistence on fixing a definite moment for the founding flawed both his accounts. By specifying either 1662 or 1664–1665 as decisive, Huet masked the process through which the academy developed. Both accounts claim to explain the founding,

[54]The one exception to this rule is found in the *Commentarius*' claim (p. 219) that Henry Oldenburg influenced Huet in the decision to create a separate academy. Huet did not know Oldenburg in 1662 and in fact had no correspondence with him until 1668, when Henri Justel arranged the first exchange.

[55]Huet to Oldenburg, 20 October 1670, *CHO,* 7:206–207.

[56]*Commentarius,* p. 219: "Nova igitur apud me instituta est Academia, quae a parvis exorta initiis, continuisque in dies aucta incrementis, parem se splendidioribus tulit."

but neither actually does so. Taken together, however, they become invaluable sources on the academy's early history. On two issues, in fact, these sources yield particularly valuable information. The first of these issues is Huet's own role in the organization; the second, the group's activities during its early, informal stage.

In the *Commentarius* and *Origines* Huet reveals a great deal about his own relation to the Académie de Physique. He clearly distinguishes between his involvement with the early group and his role in the later, more formal organization: He describes himself as patron of the early *assemblée,* as host to the later academy. Coming from Huet himself, such descriptions take on special significance. His own accounts clearly indicate that his relation to the group changed following the first stage. It is also clear that he considered his own role a limited one during both periods. Neither account claims that Huet was ever the moving force behind the scientific activity at his house.[57] On the issue of Huet's involvement with the Académie de Physique, his own testimony offers a unique source. The characterizations of his role are borne out by the contemporary evidence, but nowhere is his limited commitment to the academy described any more clearly or with more authority than in his own words. Huet considered himself the organization's first patron, then its host—never its premier scientist.

Comparison of the *Commentarius* and the *Origines* also offers a second insight into the early history of the Académie de Physique. Indeed, these accounts offer a key that unlocks the secret of the academy's early history. Implicitly, these two accounts identify the early *assemblée* as a distinct stage in the academy's development. We cannot dismiss lightly Huet's ability to draw that distinction. The only reasonable explanation for the existence of two such interpretations is that the early *assemblée* truly had "meager beginnings," and that when it "later flourished," it changed dramatically. In other words, the *Commentarius* and *Origines* suggest that the early organizational development of the Académie de Physique was not a smooth process of elaboration; instead, they indicate an initial stage that lasted three years, followed by a very rapid reorganization from which an "academy" emerged. Nothing in the contemporary documents militates against that suggestion.

One of the truly curious features of the documentation on the academy's early history is the lack of any references to organized

[57]In the *Commentarius* Graindorge's initiative created the academy; in the *Origines*, the initiative came from a larger group of *curieux*.

scientific activity at Huet's house before 1665. All references to the organized activities date from later periods. Moreover, despite the survival of voluminous quantities of Huet's letters from the early 1660s, his correspondence from that period shows no trace of significant scientific activity before 1665. In light of the claims Huet made about the Académie de Physique in his autobiography, this situation is puzzling—until we recognize the significance of the story he told in the *Origines*.

The *Origines* tells us that the scientific activity at his house began to crystallize into an "academy" only during 1665, an assertion corroborated by Graindorge's letters. The only way to reconcile the *Commentarius* version of the founding with either the *Origines* or the contemporary documents, therefore, is to conclude that the autobiography describes an early, very informal stage of development. Any group meeting at Huet's house before 1665 had to constitute an *assemblée*—something more than a scientific salon but much less than an academy.

As late as May 1665 André Graindorge referred to the scientific activity at Huet's house simply as "conversations philosophiques."[58] In September of that same year he balked at calling the activity at Huet's house "une académie."[59] Taken together with the *Origines*, such remarks indicate that we must discount the claim for a 1662 "founding" as exaggerated.

Nevertheless, Graindorge's letters, Huet's papers, and the *Commentarius* all agree that there was some scientific activity at Huet's house before 1665. More specifically, these various sources show that Huet and Graindorge had at least occasionally done dissections, most likely with the help of a paid anatomical demonstrator.[60] According to Graindorge, news of their dissecting activity had even reached Paris by the spring of 1665.[61] Along with Huet's claim (made to Henry Oldenburg in 1670) that the academy had been meeting for "eight years," these indications of an early anatomy program suggest we must accept the *Commentarius* as describing the "origins" of the anatomizing at Huet's house.

The *Commentarius*, then, gives us a significant fact. It explicitly dates the beginning of the informal activity—the *assemblée*—which later

[58]VDLI, pp. 267–269 (9 May 1665).
[59]VDLI, pp. 319–322 (16 September 1665).
[60]*Commentarius*, p. 219.
[61]VDLI, pp. 267–269 (9 May 1665).

developed into the Académie de Physique. It tells us that this preliminary stage began no earlier than 1662, and that it lasted just three years.

HUET'S *ASSEMBLÉE*

Reconciliation of the *Commentarius* and the *Origines* with the fragmentary contemporary documents yields a surprising characterization of Huet's early *assemblée*. Although this proto-organization certainly became the basis for the Académie de Physique, only hindsight allowed Huet to describe its activities as the "founding" of an academy. From all indications, Huet and Graindorge did begin dissecting in 1662. Yet nothing hints that they did so with the intention of creating a formal scientific organization. Certainly nothing suggests that they planned to form the kind of highly specialized research society they later created. On the contrary, the lack of documentation on early activity and Graindorge's obvious delight as such activity began in 1665 both argue that the *assemblée* was an amateur group that existed on an extremely informal basis—a group without any real pretensions to status as a legitimate scientific academy.

Moreover, all evidence points to the conclusion that before 1665 dissections were done only occasionally, as "rare and curious" specimens presented themselves.[62] Such dissections were almost surely conceived as demonstrations, or as the basis for *conversations philosophiques,* rather than as scientific research. Certainly any anatomizing at Huet's *assemblée* was done unsystematically, with little concern for proper laboratory technique or "modern" anatomical knowledge. Participation was only loosely organized, and undoubtedly varied each time someone found a "rare and curious" specimen for the dissecting table.

Huet's *Commentarius* and *Origines* reveal a sharp break between the stages they describe. They suggest that the reorganization of 1665 must have been dramatic and far-reaching. In short, they indicate that we cannot explain the appearance of a formal organization in

[62]The "opportunistic" nature of the academy's dissections is a phenomenon we can see throughout Graindorge's letters in the years from 1665 to 1672. Nevertheless, in mid-1665 Graindorge's surprise at what could be accomplished through systematic dissection leads us to conclude that earlier dissecting activities were even more haphazard and chaotic than those of later periods.

1665 as a natural expansion from previous activity. That being the case, we must seek the causes of the reorganization outside the routine of the early *assemblée*. Something extraordinary must have happened in 1665.

Pierre-Daniel Huet himself offers the suggestion of one such extraordinary event in the *Origines* when he attributes the formalization of the academy to the comet of 1664–1665. The comet certainly influenced Huet and apparently played an important part in changing his attitude toward the *assemblée*. Nevertheless, as with other issues, it seems Huet's emphasis on the significance of the comet resulted from his efforts to simplify and streamline the history of the Académie de Physique for the *Commentarius* and *Origines*. The comet of 1664–1665 stimulated a wave of new interest in astronomy throughout Europe, but its appearance did not precipitate the expansion and formalization of Huet's *assemblée*.

Observation of the comet of 1664–1665 influenced Huet (and ultimately the formalization of the Académie de Physique) only in the larger context of two other events. First, while the comet was visible, Huet was trying to complete the last steps in preparing his Origen edition. The comet not only fascinated him but also offered a diversion from the Origen project. In the spring of 1665 the comet made him think about abandoning Origen and devoting himself to natural philosophy. Second, in May 1665 André Graindorge had to make a trip to Paris to attend to some legal business. Almost as soon as he arrived in the French capital he discovered the scientific *assemblée* hosted by Melchisédec Thévenot.[63] The sight of this group in action fired Graindorge's imagination with the possibilities of what Huet's group might become. These two events—Huet's arrival at the final stages of the Origen project and Graindorge's discovery of the Thévenot—led to the reorganization and formalization of Huet's *assemblée*.

[63]Graindorge first attended the Thévenot on 5 May 1665. His report on this session is contained in his letter of 9 May 1665 (VDLI, pp. 267–269).

2

The *Assemblée* Becomes an Academy

André Graindorge spent eleven months in Paris, from May 1665 to April 1666. During this time he frequented various private-patronage circles and became a regular participant in the scientific academy hosted by Melchisédec Thévenot.[1] This experience proved crucial to the Académie de Physique's development. Graindorge learned a great deal at the Thévenot, and what he learned convinced him that Huet's *assemblée* could become an important scientific research society. Of equal importance, even before Graindorge returned to Caen, his knowledge of Parisian science began to change Huet's *assemblée*. In the course of their correspondence Graindorge persuaded Huet to adopt his own new approach toward science. Huet then launched the Académie de Physique de Caen as a scientific organization whose mission and purpose was spelled out in Graindorge's letters from Paris. Those letters gave Huet the means to create a true scientific academy.

Graindorge's year in Paris thus proved to be the most significant event in the history of the Académie de Physique before its royal

[1]Accounts of the Thévenot can be found throughout the literature on the early history of the Académie Royale des Sciences. That literature includes Bigourdan, *Premières Sociétés savants;* Brown, *Scientific Organizations;* Gauja, "L'Académie Royale" and "Origines"; George, "Genesis"; Hahn, *Anatomy of a Scientific Institution;* Hirschfield, *Académie Royale;* James E. King, *Science and Rationalism in the Government of Louis XIV, 1661–1683* (New York, 1972 [1949]); Martha Ornstein, *The Role of Scientific Societies in the Seventeenth Century*, 2d ed. (London, 1963); Claire Salomon-Bayet, *L'Institution de la science et l'expérience du vivant* (Paris, 1978); Joseph Schiller, "Les Laboratoires d'anatomie et de botanique à l'Académie des Sciences au XVIIe siècle," *RHS* 17 (1964): 97–114; Taton, *Origines de l'Académie Royale.*

incorporation. Despite that fact, neither Graindorge's trip nor his correspondence with Huet was planned to transform scientific activity in Caen. Graindorge went to Paris to pursue a legal case; Huet, who was at home in Caen, was determined to finish the Origen project during 1665.[2] Moreover, the Graindorge who arrived in Paris was a traditional natural philosopher—not a research scientist. He was more accustomed to *conversations philosophiques,* occasional demonstrations to buttress verbal discourses, and philological efforts to reconcile observations with texts than to systematic empirical work in a laboratory. In short, in the late spring of 1665 neither Huet nor Graindorge contemplated reorganization of the informal *assemblée.* At that point neither had the ability to plan such a change. Far from entertaining any such notion, in fact, Graindorge thought Huet's scientific *assemblée* was moribund.[3]

When Graindorge's legal business stalled in the royal bureaucracy, he began to attend the private-patronage academy hosted by Melchisédec Thévenot.[4] He then opened a correspondence with Huet, telling his patron what he had seen there. Very quickly Huet responded with questions about Graindorge's reports. Within a month their give-and-take settled into a weekly correspondence. Neither man could have entered that correspondence with the slightest notion of what it could bring. Yet their first exchange of letters in May 1665 initiated a year-long discussion of Parisian science. The correspondence changed their individual interests in science, altered their personal relationship, and recast scientific practice in Caen. Within a year they established a scientific research institute; within three years they turned that institute into a royal academy of sciences. By any measure, then, Graindorge's year in Paris had remarkable consequences. In effect, his letters brought a scientific revolution to Caen.

[2]Although Graindorge never specified the nature of his legal business in his letters to Huet (presumably Huet was familiar with the case), he did make frequent references to his lack of progress with it. The most explicit and detailed presentation of the legal procedures involved can be found in his letter of 10 October 1665 (BL, A 1866, 1982). Apparently he was petitioning for some personal right or exemption based on his title of nobility.

[3]VDLI, pp. 267–269 (9 May 1665). In this letter Graindorge asks to be told "si vous ne continuez pas la conversation philosophique."

[4]Different sources offer various dates for the closing of the Thévenot. Brown puts that academy's demise "about 1665" (*Scientific Organizations,* p. 136), George "in the fall of 1664" ("Genesis," p. 375); and Hirschfield asserts that the Thévenot closed "before June 1665" (Académie Royale, p. 12).

THE MECHANICS OF CHANGE

In the spring of 1665 Huet faced a personal crisis. After putting the Origen manuscript aside long enough to write a treatise on the comet of 1664–1665, he found himself extremely reluctant to return to work as a translator and exegete.[5] He had come to doubt both his ability to finish the Origen project and its value should he bring it to completion. In March he signaled these doubts to his patron, Jean Chapelain. Huet had reached a level of frustration where he considered Origen "bas & obscur" and thought his real talents might lie with astronomy and natural philosophy.[6] Apparently he sought Chapelain's approval for a decision to abandon the Origen project. Certainly he solicited advice on whether he should devote more time to natural philosophy. Chapelain's response was sharp and to the point. He insisted Huet return to the Origen manuscript, telling him he could do nothing better for the public welfare than complete this work, which was capable of "strengthening Christians in the faith and destroying the machines this impiety has raised up against it." Chapelain then advised that while completing the *Origenis Commentaria*, Huet should "lend" his talents to natural philosophy only when nature offered some marvel worthy of his attention.[7]

Huet followed his patron's advice. He returned to the Origen project, and by the time Graindorge left Caen (early May 1665), he had become so involved with the work that Graindorge feared he might give up natural philosophy altogether.[8] By April 1666, when

[5]For analysis of Huet's treatise on the comet, see Tolmer, *Huet*, pp. 291–300.

[6]In Huet's papers, we find three versions of Huet's treatise on comets (BL, A 1866, 2408, 2409, and 2505). All are addressed to Chapelain; all are dated 5 March 1665. Clearly these three versions represent successive drafts of the same work. In the final draft (2505), Huet simply introduced his subject by saying, "J'aurais pris plaisir à signaler mon art astronomique." Apparently that statement alerted Chapelain that Huet had lost enthusiasm for the Origen project and provoked the admonition discussed below. In the first draft (2408), Huet had been much more explicit about both his frustrations and his intentions: "Il est vrai qu'une nouvelle traduction [of the Origen manuscript] seroit fort necessaire, mais . . . j'employe mon tems si peu volontiers a traduire ce travail . . . bas & obscur, que je suis resolu d'entreprendre a l'avenir le moins de traduction que ie pourrais. . . . Croyez-vous, Monsieur, qu'il m'est plus agreable d'examiner ces diverses observations de la dernier comete. . . ."

[7]Chapelain to Huet, 14 March 1665, BL, A 1866, 255: "Il faut vous louer de vos inclinations philosophiques qui vous font tourner les yeux du coste de la nature et les attachent a la contemplation des merveilles quelles produit de temps en temps."

[8]VDLI, pp. 267–269 (9 May 1665).

Graindorge returned from Paris, Huet had the completed manuscript for the *Origenis Commentaria* ready for the printer. In fact, Huet left Caen to arrange its publication on the very day Graindorge arrived home from Paris.[9]

Graindorge knew about Huet's determination to finish the Origen project during 1665; nevertheless, his patron's rapid progress with the work surprised him. Huet's continuing interest in natural philosophy surprised him even more, however. Graindorge left for Paris thinking his patron might abandon *la conversation philosophique* and was thus pleased to learn that Huet continued dissecting. At midsummer he remarked at Huet's tenacity in pursuing both his dissections and the work on Origen.[10] In the fall (after Huet had a complete draft of his Origen manuscript), Huet's new series of systematic dissections elicited excited commentaries from Graindorge. In January 1666 Huet's successes with the "new" anatomy program earned effusive praise. By March, Graindorge was talking about the "sharp new start" Huet had made with the *assemblée* and worrying about the effects Huet's coming departure from Caen (to find a printer for his work) might have on the "Académie de Physique de Caen."[11] When Graindorge returned to Caen in April 1666, Huet had just left for Rouen, but he had left an academy that followed a defined program, that had a corps of regular members, and that met every Thursday evening.

In a sense, Huet had merely followed Chapelain's advice. He had completed the *Origenis Commentaria* and while doing so had "turned his eyes to the contemplation of nature" and seized a rare opportunity. Indeed, with his *fidèle* acting as "intelligencer," Huet found himself in a unique situation—one in which he could hardly complain about the need to shift his scientific focus from astronomy to anatomy and dissection. Using Graindorge's letters, he had restructured his scientific *assemblée*. He had created an academy. The significance of Graindorge's letters had reached far beyond the transmission of a body of scientific knowledge. They had transmitted a new set of social relations. In creating an academy, Huet depended on those letters. The degree of his dependence becomes apparent as we examine the mechanics of the Huet–Graindorge correspondence.

Graindorge was in Paris for forty-nine weeks.[12] During that time he

[9]BL, A 1866, 541 (10 April 1666).

[10]VDLI, pp. 267–272, 300–322 (9 and 19 May, 29 July, and 16 September 1665).

[11]BL, A 1866, 551, 639, 648 (9 December 1665, 20 January, and 10 March 1666).

[12]His correspondence spanned forty-eight weeks. His first letter (9 May 1665) reveals, however, that he had attended the previous Tuesday's session at the Thévenot

Table 1. Graindorge's 45 letters from Paris, 9 May 1665–10 April 1666, by week,* date, and day written

*Week**	*Date*	*Day*	*Week**	*Date*	*Day*
1	9 May	Saturday	26	—	
2	—		27	4 November	Wednesday
3	19 May	Tuesday	28	11 November	Wednesday
4	30 May	Saturday	29	18 November	Wednesday
5	3 June	Wednesday	30	25 November	Wednesday
6	10 June	Wednesday	31	2 December	Wednesday
7	17 June	Wednesday	32	9 December	Wednesday
8	24 June	Wednesday	33	16 December	Wednesday
9	1 July	Wednesday	34	23 December	Wednesday
10	8 July	Wednesday	35	—	
11	15 July	Wednesday	36	5 January	Tuesday
12	22 July	Wednesday	37	13 January	Wednesday
13	29 July	Wednesday	38	20 January	Wednesday
14	5 August	Wednesday	39	27 January	Wednesday
15	13 August	Thursday	40	3 February	Wednesday
16	19 August	Wednesday	41	9 February	Tuesday
17	—		42	17 February	Wednesday
18	2 September	Wednesday	43	24 February	Wednesday
19	9 September	Wednesday	44	2 March	Tuesday
20	16 September	Wednesday	45	10 March	Wednesday
21	23 September	Wednesday	46	17 March	Wednesday
22	3 October	Saturday	47	22 March	Monday
23	10 October	Saturday		24 March	Wednesday
24	—		48	31 March	Wednesday
25	24 October	Saturday	49	10 April	Saturday

*Weeks (Monday through Sunday) counted from 4 May 1665.

addressed forty-five letters to Huet. He dated every letter, and when we arrange them chronologically (see Table 1), two patterns emerge. First, they span the entire forty-nine weeks with remarkable regularity. Seldom did Graindorge allow a week to pass without writing.[13] For long periods his letters followed at regular weekly intervals. By itself, the consistency of this correspondence over such a long time is extraordinary—unrivaled in any other period of the Huet–Graindorge correspondence.

The second pattern appears as we match the dates of Graindorge's

(5 May 1665). He arrived in Paris before that date. His last letter (10 April 1666) speaks of his intention to leave Paris the next day (11 April 1666).

[13]When we count the weeks from Monday to Sunday (counting backward from his stated date of departure, Sunday, 11 April 1666), we find he failed to write a weekly letter only five times in the entire course of the correspondence (see Table 1).

letters with the daily calendars for 1665 and 1666. The mail left Paris for Caen twice a week (Wednesday and Saturday),[14] but Graindorge had a definite preference for the Wednesday mail. He mailed a letter to Huet on Wednesday in thirty-seven of the forty-nine weeks he was in Paris.[15] During two long periods (June through September 1665; November 1665 through March 1666) the midweek pattern remained virtually unbroken. In the second of those periods (November through March) he missed sending a Wednesday letter only once in five months (twenty-two weeks).

Graindorge was conscious of his letter-writing habits. Whenever he broke the Wednesday pattern, his next letter offered some comment on missing his *jour ordinaire.*[16] He clearly wanted to send his weekly letter, but even more important, he wanted that letter to leave Paris on Wednesday. The explanation for such a conscientious habit is in part simple. The schedule of the Thévenot academy, which met on Tuesday evening, set Wednesday morning as the logical time to write Huet, and through September 1665 his letters concentrate almost exclusively on reports of activities at the Thévenot. Nevertheless, although the schedule of the Thévenot can explain the desirability of Wednesday-morning letter writing, it cannot explain why he switched to this system only after a full month in Paris, nor can it explain why he held to the routine so faithfully during two extremely long periods. Why did he switch to Wednesday? And then why did he maintain that schedule so faithfully?[17] To answer those questions, we must see how the mail could influence the sessions of Huet's *assemblée*.

Graindorge's letters reveal three essential pieces of information about mail service between Paris and Caen. First, he received mail in Paris twice each week—once on Monday, again on Saturday.[18] Sec-

[14]VDLI, pp. 272–274 (30 May 1665). For details on the postal system, see Eugène Vaillé, *Histoire générale des postes française,* vol. 3, *De la réforme de Louis XIII à la surintendance générale des postes (1630–1688)* (Paris, 1950).

[15]It seems reasonable to assume that he used the Wednesday mail for all letters written from Monday through Wednesday.

[16]VDLI, pp. 305–308 (13 August 1665): "Je ne sais par quel désordre vos lettres ont changé de jour"; BL, A 1866, 1980 (3 October 1665): "Je ne scais pas bien si fut paresse ou affaires qui mempescherent de vous escrire mon jour ordinaire"; BL, A 1866, 1982 (10 October 1665): "Je ne scais pourquoy Jay changé le jour ordinaire"; BL, A 1866, 546 (24 October 1665): "Je voulois repondre mon jour ordinaire"; BL, A 1866, 549 (5 January 1666): "Vous mavez donné le pion en me chargeant de deux de vos lettres." The consistency of these comments confirms that the forty-five letters listed in Table 1 constitute a complete series.

[17]For an explanation of his lapse in October, see n. 19 below.

[18]VDLI, pp. 272–274 (30 May 1665).

ond, as Graindorge posted a letter in Paris on any given Wednesday, he expected a reply to the same letter on the following Monday. Third, Graindorge's letters demonstrate that Huet's half of the correspondence frequently made reference to new research undertaken in Caen (and that Graindorge was reading Huet's letters at the Thévenot). When we take these three pieces of information together, it becomes obvious that the letters Graindorge put into the mail late Wednesday morning reached Caen in time to be read (and acted on) at Huet's *assemblée* that same week. As the mail arrived in Caen on Monday and Thursday, Huet received Graindorge's Wednesday letters from Paris late the next day. In sum, then, Graindorge posted his letters on Wednesday morning because he knew they would reach Huet in time for a scientific *assemblée* on Thursday evening.

During the summer of 1665 the following pattern established itself. Graindorge attended the Tuesday-evening session of the Thévenot in Paris. Early Wednesday morning he wrote his letter to Huet. Thursday evening Huet read Graindorge's letter to the group gathered at his house in Caen. In turn, Huet wrote to Graindorge reporting on demonstrations and discussions at his own *assemblée*. That letter left Caen on Saturday and reached Paris on Monday. Graindorge reported its contents at the Thévenot on Tuesday. The cycle then repeated itself. In that way, as long as Huet and Graindorge kept up their respective halves of the correspondence, Huet's provincial *assemblée* maintained weekly, session-for-session communications with the Thévenot. Little wonder Graindorge put his letters in the Wednesday mail so faithfully.[19]

Recognition of the cycle of communications implied in the Huet–Graindorge correspondence provides a series of insights into the effects of Graindorge's year in Paris. This extraordinary organizational dialogue with the Thévenot explains why Huet organized his *assemblée* around weekly sessions during 1665.[20] Participation in such an exchange presented an irresistible opportunity, but it was one that

[19]An understanding of how the cycle of communications worked between Paris and Caen explains the one significant breakdown in the cycle, which occurred in October 1665 (see Table 1). During this period, Graindorge described the Thévenot as "en vacances" (BL, A 1866, 546, 644 [24 October, 4 November 1665]).

[20]To cite an interesting parallel, Moisant de Brieux explicitly stated that the organization and meeting time for his Académie du Grand Cheval had been fixed by the arrival of the mail on Monday afternoon. His account, which was originally contained in the preface to a volume of his verses, has been reprinted by François-Richard de La Londe in his *Mémoire pour servir à l'histoire de l'Académie Royale des Belles-Lettres de Caen* (Caen, 1854), pp. 9–11.

Huet could exploit only if his group met every week. Likewise, the possibility of weekly exchanges on new research explains why Huet committed himself to dissection rather than to astronomy while Graindorge was in Paris. He simply adopted the program offered in Graindorge's letters. Just as surely, this pattern of exchanges with the Thévenot also explains why Huet chose Thursday evening as the time for his sessions.[21] Finally, once we recognize this weekly pattern in the Huet–Graindorge correspondence, Huet's ability to generate new interest and participation in his *assemblée* becomes easily comprehensible. In fact, by maintaining his *jour ordinaire* so faithfully, Graindorge ensured that his letters played a major role in changing both the organizational form and the level of scientific activity at Huet's house.

Graindorge's letters themselves indicate that the level of group activity at Huet's house changed dramatically between May 1665 and March 1666. Graindorge's first reference to Huet's *assemblée* appears in the complimentary close of his letter of 3 June 1665: "I send greetings to all our gentlemen of the one and of the other academy." Since the only other "academy" Huet could have been attending besides the Grand Cheval would have been the one meeting in his own house, this was certainly a reference to those sessions. Despite that reference to an "academy," however, it is clear that Huet's activities still did not draw a regular group. Just a few weeks later Graindorge wrote: "I am sorry that you find yourself alone at your dissections. Some do not want to take the trouble, and others consider it only caprice, but it is important never to give up."[22] Taken together, these two references to group activity indicate that Graindorge wanted an "academy," but that none yet existed in Caen.

The status of Huet's *assemblée* remained in question in September 1665, when Graindorge next referred to organized activity at Huet's house by specifying it was a "society, to which I will not dare give the name academy, out of fear of creating jealousy at the Grand Cheval."[23] Graindorge's hesitancy about calling Huet's *assemblée* "une académie" may have stemmed from a fear of hostility from the Grand Cheval, as he claimed, but more likely it revealed his doubts about Huet's commitment to regular sessions. In either case, Graindorge

[21] Ibid. Undoubtedly the fact that the Grand Cheval was already meeting on Monday afternoon also helped to make Thursday (and the Thursday mail) attractive to Huet.

[22] VDLI, pp. 275–278, 288–290 (3 June and 1 July 1665).

[23] VDLI, pp. 319–322 (19 September 1665). This "fear" of the Grand Cheval reappears in Graindorge's letter of 23 December 1665 (BL, A 1866, 638).

still did not perceive Huet's *assemblée* as a legitimate academy as late as September 1665.

By early 1666, however, Graindorge's hesitancy disappeared, and he began to use the title "Académie physique de Caen." In his letter of 22 March 1666, for example, Graindorge once again constructed his complimentary close to address readers other than Huet. This time, however, he did not equivocate: "I greet our gentlemen of the Académie physique."[24] Between September 1665 and March 1666, then, Graindorge had gradually concluded that the activity at Huet's house warranted the term "academy."

Although it is difficult to date the start of actual group activity, Graindorge's letters indicate that Huet's *assemblée* began to attract regular attendance in November and December 1665—following the start of the second long period during which Graindorge wrote his letters on his *jour ordinaire*.[25] His letters from that time demonstrate that Huet had all but finished the final draft of the Origen manuscript, that he was devoting more time than ever to anatomy, and that at Graindorge's urging he had begun a new program of dissections.[26] From all appearances, in fact, during this period Huet began to take his science more seriously than ever before. Graindorge's letters from late 1665, for example, show that during this period Huet began to argue the importance of mathematics as the key to philosophy and metaphysics. By early 1666 Huet was working on a series of very specific scientific questions with his dissections, and presumably the sessions of his *assemblée* had settled into a regular pattern.

Graindorge did not comment on the increase in activity or the regularization of the sessions until these phenomena were well established. Nor did he ever intimate that his own letters might have played a rôle in changing Huet's *assemblée*. Nevertheless, when Graindorge finally began to describe the activity at Huet's house as an "académie," the situation in Caen had changed dramatically from what it had been just a few months earlier. Graindorge's letter of 10 March 1666, for example, not only takes the existence of an organized society for

[24]BL, A 1866, 662 (22 March 1666).

[25]During this period Graindorge made only one comment that indicates Huet was making structural changes in the *assemblée*. In his letter of 23 December 1665 (BL, A 1866, 638) he described Huet's "academy" as a group "qui change ce me semble bien souvent de jour." Suggestive as that statement may be, it is too ambiguous to help us date the start of group activity.

[26]VDLI, pp. 319–322 (16 September 1665); BL, A 1866, 546 (24 October 1666). In the second letter Graindorge apologized for failing to respond promptly to various of Huet's queries in a series of letters that had "piled up."

granted but also expresses the hope that the academy can continue along its newfound path. When Huet told Graindorge he was leaving Caen to find a printer for the *Origenis Commentaria,* that news prompted Graindorge to respond: "I worry about the Académie physique de Caen because, without you, I see no one so well loved, so well intentioned, nor so dedicated. You have made a sharp new start there, and you have advanced in the path we marked out. In the old days, we produced nothing."[27] In contrast with Graindorge's remarks in the summer and fall of 1665, this passage shows that by the early spring of 1666 the Académie de Physique had assumed the characteristics of a formal, albeit embryonic, scientific organization. Graindorge no longer worried whether Huet would continue the sessions. He no longer even doubted he could call the group an academy. By March 1666 Graindorge worried about maintaining the "sharp new start" during Huet's absence. In other words, Graindorge no longer saw the problem as how to establish the academy; the problem had become how to maintain the one that existed.

Because Huet organized the sessions of his *assemblée* around Graindorge's letters, the intellectual content of those letters gained unusual force. While Graindorge attended the Thévenot, his letters led Huet to structure his own research program around topics current there. Very quickly, in fact, Huet's use of Graindorge's letters transformed them into the official statements on topics and methods the Académie de Physique would pursue. Thus Graindorge's ideas on science took on unique importance. As long as he was Huet's intelligencer in Paris, Graindorge exercised a dominating influence over the type of science practiced in Caen. In effect, he became Huet's expert adviser. Once that happened, Graindorge controlled the academy's scientific program. For that reason, the nature of the changes in Graindorge's own approach to nature during his year in Paris is central to an understanding of the research agenda established by the Académie de Physique.

A PROVINCIAL *CURIEUX* IN PARIS

André Graindorge's letters show a distinctive conception of scientific activity at the time he arrived in Paris. He perceived a hierarchy of merit among the practitioners of natural philosophy and thought

[27] BL, A 1866, 648 (10 March 1666).

that there was a corresponding hierarchy in the topics available for research. He thought these "dignities" of the researcher and his research should be closely matched. According to Graindorge, the great minds of the age should concern themselves with great issues—not with mundane or undignified topics. For example, he did not expect a savant of Huet's stature to involve himself with the messy details of preparing specimens for dissection.[28] Instead, he thought someone like Huet should concentrate on those issues "worthy" (*digne*) of his efforts.[29] By extension, this logic led Graindorge to think that expert anatomists (such as Huet or himself) wasted time in dealing with such lowly creatures as frogs, rabbits, and fish.

What Graindorge saw in Paris immediately challenged such ideas. In his first letter to Huet, Graindorge described a session of the Thévenot academy at which the greatest anatomist of the day, Niels Steno, had dissected a rabbit.[30] Graindorge was amazed at what he saw—it appears that he had never before even seen the intestines of a rabbit.[31] Despite his claims to expertise, then, it is clear that his knowledge of vertebrate anatomy was actually rudimentary. For Graindorge, watching Steno dissect was like having an entire new

[28]In response to Huet's report that he had dissected a dog, Graindorge said he was glad Huet had finally become so "inured" to the suffering of animals (VDLI, pp. 269–272 [19 May 1665]). The obvious implication is that until this time Huet had had little to do with the preparation of specimens.

[29]When he arrived in Paris, Graindorge thought he and Huet were already accomplished anatomists. Though the actual quality of their work is impossible to judge from the surviving documents, Graindorge's attitude toward his own expertise can be seen in his initial willingness to compare his and Huet's dissections favorably with those done in Paris. In his first letter to Huet (VDLI, pp. 267–269 [9 May 1665]) he claimed: "L'on sacvait que nous étions mis pareillement sur la dissection à Caen, et Mr. Thévenot nous exhorta fort à disséquer toutes sortes de poissons dont ils sont privés. . . . Nos dissections du cerveau en général ont pour le moins aussi bien réussi que celles que je vis."

[30]Niels Steno (1631–1686) was born in Copenhagen and studied medicine at the university there. In 1660 he went to Amsterdam and then to Leiden to study anatomy. It was during his time in the Low Countries that he discovered the parotid duct, which is also known as Stensen's duct. In 1664 he came to Paris under the patronage of Thévenot. In September 1665 Steno left Paris for Italy, where he joined the group at the Cimento. In 1667 he converted to Catholicism. In 1676 he was named bishop of Titiopolis and appointed apostolic vicar of northern Germany and Sweden.

[31]VDLI, pp. 267–269 (9 May 1665): "L'intestin coecum est une pièce rare et monstrueuse dans les lapins." Nor was his lack of knowledge about anatomy limited to unfamiliarity with rabbits. In his letter of 3 June 1665 (ibid., pp. 275–278) Graindorge said he thought frogs worthless for demonstrating how the blood circulates because "l'on m'a dit que ce [the heart] n'est qu'une pellicule blanche qui rougit à mesure que le sang y entre." Just a week later he reversed himself—apparently after seeing the dissection of a frog for the first time: "Le coeur de la grenouille . . . est admirable pour y voir entrer le sang" (ibid., pp. 278–281 [10 June 1665]).

world opened up before his eyes. Not only did the Dane seem willing to cut open anything and everything put before him but he was also constantly at work at his dissecting table. The effect on Graindorge was immediate, and by the time he wrote his second letter to Huet, his praise for Steno knew almost no bounds:

> This M. Steno is the rage here. This evening after dinner we saw [him dissect] the eye of a horse. To tell you the truth, we are only apprentices next to him. I begged him to show me a heart tomorrow morning, which, with singular goodwill, he promised to do. He is constantly dissecting. He has patience that is inconceivable, and with practice he has acquired a technique above the ordinary. Neither a butterfly nor a fly escapes his skill. He would count the bones in a flea—if fleas have bones.[32]

Such enthusiasm can only mean that the program in Caen did not measure up to this standard. Graindorge himself admitted as much in saying that he and Huet rated only as "apprentices" by comparison to Steno. Overall, in fact, the enthusiasm for the Thévenot we find in Graindorge's early letters from Paris leads to the unavoidable conclusion that the science he was familiar with at Huet's *assemblée* was rudimentary in comparison with the Parisian standard.[33]

Graindorge had arrived in Paris thinking that he and Huet were experienced anatomists who should deal only with research "worthy" of their talents. Apparently Huet shared that notion, because his response to being called an "apprentice" prompted Graindorge to spell out exactly why he had chosen that derogatory term. He had learned any number of new things in just a few weeks of watching Steno. He refused to back down:

> When I called us apprentices next to M. Steno, I had reason, for I have never seen such dexterity. He made us see everything there is to see in the construction of the eye—without putting either the eye, the scissors, or his one other small instrument anywhere but on his one hand, which he kept constantly exposed to the gathered company.
>
> I would be very pleased if you lifted the cornea without touching the uvea. That is the very thing he did so well, which I was seeing for the

[32]VDLI, pp. 269–272 (19 May 1665).

[33]For an introduction to contemporary practices, see Maurice Caullery, "La Biologie au XVIIe siècle," *XVIIe Siècle*, no. 30 (1956): 25–45; Jacques Roger, "Réflexions sur l'histoire de la biologie (XVIIe–XVIIIe siècle): Problèmes de méthodes," *RHS* 17 (1964): 25–40.

> first time. Moreover, I saw, but in the head of a cow [rather than in the horse from which he took the eye], certain glands above the eye, which furnish tears through small branches open to the eyelid. I saw his passage *salivaire exterieur.*[34] I saw the course of the fibers in the heart as well as the part they play in the construction of muscles.[35]

Graindorge was indeed learning new things in Paris, and as he sent his letters to Caen, Huet quickly became interested in finding out more about Parisian science. In the first week of June, Graindorge began to write his weekly letters on his *jour ordinaire,* and Huet undertook new research based on the information Graindorge reported. By mid-June, for example, Huet was doing vivisections of frogs, trying to demonstrate the circulation of the blood, and working to find the cause of the "blind spot" optical illusion.[36] Each of these new projects had been inspired by news Graindorge had sent from Paris.

Despite his enthusiasm for Parisian science, Graindorge did not immediately abandon his conception of science as a hierarchical activity in which the a priori dignities of researcher and research needed

[34]Stensen's duct.

[35]VDLI, pp. 272–274 (30 May 1665).

[36]Ibid., pp. 269–272, 275–285 (19 May; 3, 10, and 17 June 1665). In the three June letters Graindorge offered a series of remarks on Huet's new work with the circulatory system, for which he was apparently using frogs as his laboratory specimens. Graindorge's remarks indicate that Huet claimed his frogs continued to live after their hearts were "removed." The most logical explanation for such a claim is that Huet was actually doing vivisections. In short, he was keeping the frogs alive with their hearts "exposed." It seems Huet opened this line of research following on just a single remark that Graindorge made in his first letter from Paris: "Je vis le scelet d'une grenouille fort joliment préparé. Quelqu'un me di qu'étant privée de son coeur el vécut encore fort longtemps" (VDLI, pp. 267–269 [9 May 1665]). Despite Hirschfield's claims (*Académie Royale,* pp. 136–146) that Mariotte first discovered the blind spot, its existence was clearly understood before Mariotte's work in 1668. During May 1665, in fact, the phenomenon was introduced at the Thévenot along with a challenge to the academicians to explain the cause. Graindorge did not name the person who issued this challenge, but he did pass it along to Huet. Over the next month, Huet did various trials and was able to explain the blind spot as a result of insertion of the optic nerve. Graindorge's commentaries on this phenomenon and the questions of anatomy and optics raised by attempts to deal with it are far too extensive to quote or summarize here, but it is important to cite his initial statement explaining how Huet could demonstrate the illusion: "L'on nous proposait l'autre jour d'où vient que, regardant un objet en droite ligne, par exemple votre oeil *O* regarde *A*, puis regardant fixement *B*, il perd l'espèce de *A*. Ensuite, il regarde *C;* il découvre *A* qu'il avait perdu" (VDLI, pp. 272–274 [30 May 1665]). Graindorge then offered extended commentaries on both Huet's work and the discussion of this subject in his letters of 10, 17, and 24 June 1665 (pp. 278–288). For discussion of Mariotte's work with this phenomenon, see G. Bugler, "Un Précurseur de la biologie expérimentale: Edme Mariotte," *RHS* 3 (1950): 242–250.

to be matched. He was impressed by the dissections he saw Steno doing, and he saw the value in them. Nevertheless, Graindorge was not sure that this activity was *digne* for someone of Huet's stature. He though Huet should reserve his talents for truly important research. Commenting on Huet's search for the ova in a female dog, for example, Graindorge was surprised Huet had sacrificed the specimen himself but told his patron that "this is a topic worth [*digne*] of your investigation."[37]

Graindorge's perception of Huet as a gifted man of letters made him think that Huet should be equally gifted as an anatomist. That opinion survived Graindorge's first exposure to Steno's dissections; in fact, his ideas about the "dignities" of science only slowly came under attack during the summer of 1665. By the end of July, however, Graindorge was having difficulties balancing his hierarchical ordering of scientific activity against what he had seen at the Thévenot. He had not yet abandoned the notion that Huet was a superior savant who should reserve his talents for the really important intellectual tasks, but he was becoming troubled by that idea.

In his letter of 29 July, Graindorge reflected on Huet's intellectual program. The ironical juxtapositions of science and belles lettres reveal the difficulties he was having in reconciling Huet's recent dissection of a carp with his dignity as a savant:

> I am glad to hear that you are occupied with research on Origen's name as well as the research on his voyages; that says you are off the tenets [of religion found] in this little diamond of an author—those have tied you up horribly and have been an entire first labor rather than a prelude. You are fortunate to master so many wonderful things at the same time.
>
> You take Origen by the ears, you attend the academy of M. de Brieux; and yet you do not fail to examine the head of a carp. As for myself, I swear that when I ponder the charms of belles lettres, which I see posing so many wonderful questions (as well as resolving so many of the difficulties found in the good authors),[38] your pursuits are worthy of great men. Do you not agree? And do you not understand the excellence of the anatomy of a verse of Homer or Virgil—an excellence infinitely more lofty than the dissection of a worm.? The other day we carefully examined the foot of a fly [with a microscope]. . . . Would it not be worth more to examine the meter of a foot of verse?[39]

[37]VDLI, pp. 269–272 (19 May 1665).
[38]The ancients.
[39]VDLI, pp. 300–303 (29 July 1665).

In the contexts of Graindorge's relationship with Huet; his attitudes toward belles lettres, philology, and the Grand Cheval; and his enthusiasm for natural philosophy, this passage must be read as carrying a heavy load of irony.[40] It must also be read as expressing disaffection from the way he had been looking at the world. Clearly Graindorge was not expressing his own normative views on how important science *should* be; yet the irony is obviously based in his perception of the fact that his contemporaries (including Huet and himself) had given science a lower status than belles lettres. Graindorge was struggling with that arrangement of priorities.

Even the language in this letter takes on significance in conveying Graindorge's message. The metaphors Graindorge constructed to describe Huet's activities jumble the lofty with the base: "Vous tenez Origène par les oreilles . . . et vous ne laissez pas d'examiner un tête de carpe." Graindorge added to the effect with an extended play on the words *vers, ver,* and *pied* as he compared the anatomy of "un vers d'Homere" with the anatomy of "un ver de terre," and "un pied de mouch" with "un pied de vers." Beyond any question, Graindorge was playing with the contemporary perception of observational science as an activity carrying lower "dignity" than belles lettres and philology. The dissection of the carp was not "worthy," and Graindorge expressly pointed out the bizarre oddity that Huet, the savant who dealt with the "little diamond" Origen, still did something as lowly as anatomize fish.

Graindorge always retained a great deal of respect for Huet the savant, but during his stay in Paris his views on Huet's science changed. He began to draw distinctions between Huet the gifted savant and Huet the scientist. As a result, his evaluations of the scientist became

[40]Brennan, "Culture and Dependencies," pp. 144–145, interprets this letter as "demonstrating Graindorge's respect for the Academy at Moisant de Brieux's and for *belles-lettres* in general." Although a straightforward reading of Graindorge's remarks can certainly be interpreted in that vein, it hardly seems likely (given other remarks Graindorge had made to his patron) that Huet could have believed that Graindorge intended his words to be taken literally. To cite a contemporary example of Graindorge's attitude toward Brieux and the Grand Cheval, we need look no further than his letter of 3 June 1665 (VDLI, pp. 275–278), in which Graindorge commented on the fact that the Grand Cheval had not met for several weeks: "Est-ce que le mal de Mr. de Brieux a empêché qu'il n'y ait eu d'Académie, ou si c'est par pure paresse, ou bien si la physique vous fait oublier les belles-lettres? Si cela est, je conseillerais à Mr. de Brieux de ne revenir point de son mal. Je suis ravi qu'il soit tiré d'affaire." The Huet who read that statement in June can hardly have read this letter of 29 July as expressing Graindorge's deepest feelings or any real respect for the Grand Cheval.

more moderate. The way this change came about is intriguing. Rather than occurring as a result of Graindorge's new definition of science, the alteration in his perception of Huet's scientific capabilities came first, and actually may have been the primary cause of larger changes in his ideas about how science should be practiced.

Huet's lofty position in the intellectual community of provincial Caen may have exempted his scientific pronouncements from close scrutiny at his own *assemblée,* but his reputation at home did not protect his science from careful examination in Paris. As soon as Huet began to send reports of his new researches to Graindorge, Graindorge had to face the unpleasant task of telling his patron how some of them were received at the Thévenot. The necessity of this faultfinding with Huet's science could not fail to have an impact on Graindorge.

In mid-June, Graindorge responded to Huet's claim to have made an important new discovery about the anatomy of chickens with an abrupt report on how it had been dismissed at the Thévenot: "They strongly believe here that what you are calling the milk ducts in chickens are nothing but the same vessels already described. I myself think that these are the ones that carry the egg white to the egg—no one has been able to find milk ducts [in chickens]."[41] Apparently Huet did not take kindly to the rude way his discovery was dismissed at the Thévenot, because the next Wednesday Graindorge felt obliged to clarify his previous comments. Obviously trying to be conciliatory, Graindorge remained firm nonetheless: "This is not to contradict what you have seen with your own two eyes, but only to ask you to examine the truth of the fact carefully—no one has ever found milk glands in birds."[42]

Huet's report describing milk ducts in chickens was not the only one of his amazing discoveries that raised doubts at the Thévenot. Graindorge also found that Huet's account describing a frog that had lived for a week without a heart was greeted cynically. Once again he had to tell his patron the bad news: "How in the devil is that frog still living? . . . Tell us the truth because no one here dares believe it."[43]

Once again Huet felt offended, and the next week Graindorge returned to the subject: "I do not doubt the truth of your *expérience* with the frog. I have told you that, but when you say to me a week

[41]VDLI, pp. 278–281 (10 June 1665).
[42]Ibid., pp. 282–285 (17 June 1665).
[43]Ibid., pp. 279–281 (10 June 1665).

later, 'what would you say if our frog was still living?' that is what surprised me and that is what I took for confirmation [that it was living], and that was the cause of my doubt."[44]

Such requests for "refinements" in Huet's interpretations form a consistent pattern in Graindorge's letters from Paris. Graindorge was deferential, but his criticisms were pointed and direct nonetheless—he was telling his patron that his observations were mistaken and his interpretations were absurd. He still had respect for Huet the savant, but he had begun to realize that those skills did not guarantee an accurate observer, especially when interpretations were based on a single dissection.

By December 1665 Graindorge's faith in Huet's scientific abilities had begun to renew itself. He even began to have more confidence in Huet's dissections. Significantly, however, his new evaluation of Huet sprang from new assumptions.[45] Graindorge had dropped the a priori dignities of natural philosophy. In his letter of 2 December 1665, for example, Graindorge responded to Huet's announcement of plans for a new series of dissections on dogs with a comment that reveals his new standards as an anatomist: "A greyhound is excellent for dissection; its shape and thinness make it wonderful for seeing things that are difficult to find otherwise."[46] No longer did the specimen have to be *digne;* it had only to demonstrate what the anatomist wanted to see.

Huet had not only planned new dissections of greyhounds but adopted an entirely new approach to his anatomy program. The week after he first mentioned the greyhounds, Graindorge began his letter with the same topic. The context of his discussion makes it clear that Huet was not only dissecting but also going through the mundane and laborious process of checking earlier observations with this new series of dissections. Graindorge was thrilled: "I wanted to grab your greyhound by the ears [when I read your letter]. It seems to me that I will always be well served by your observations because whatever

[44]Ibid., pp. 282–285 (17 June 1665).

[45]For discussion on the development of empirical practices in seventeenth-century science, see Léon Auger, "Le R. P. Mersenne et la physique," *RHS* 2 (1948): 33–52; Charles Dedel, "La Pharmacie au XVIIe siècle," *XVIIe Siècle,* no. 30 (1956): 46–61; Luigi Belloni, "Marcello Malpighi and the Founding of Anatomical Microsopy," in *Reason, Experiment, and Mysticism in the Scientific Revolution,* ed. M. L. Righini Bonelli and William R. Shea (New York, 1975); Steven Shapin and Simon Schaffer, *Leviathan and the Air-Pump: Hobbes, Boyle, and the Experimental Life* (Princeton, 1985).

[46]BL, A 1866, 640 (2 December 1665).

pains are taken, something always escapes our notice. It is necessary to repeat one's research frequently."[47]

Discussion of Huet's dissections dominated the correspondence for the next several months. Graindorge could not find enough praise for Huet's new interest in a careful program of dissections. In his letter of 20 January 1666, for example, Graindorge again compared Huet's skill as an anatomist with that of Niels Steno. The earlier insult of referring to Huet as an apprentice had clearly not been forgotten:

> When I saw M. Steno dissect an eye, I wrote to you that we were novices. Since he carried out his dissection with such patience and dexterity I could not exaggerate his skills enough. He did almost everything you are pointing out to me except the way to see the pigments. If I had seen you dissecting the eye of this cow, I would not have described you as a novice."[48]

By January 1666, then, Graindorge had regained confidence in Huet's scientific abilities, but his confidence sprang from new criteria. Rather than seeing Huet's task as the pursuit of unique and special research, Graindorge had come to envision it as part of the scientific community's efforts to extend and confirm knowledge. He was becoming more and more convinced that the only path toward an accurate understanding of nature lay in large-scale, empirical investigations in which all *curieux* shared the same kind of work. In his mind, traditional natural philosophy was giving way to science.

Graindorge's experiences in Paris had led him to a new concept of science.[49] By the fall of 1665, what he had seen at the Thévenot and the problems he had seen in Huet's work led him to abandon the dignities of natural philosophy as a guide to organizing research. In place of that system he adopted an empiricism based on concepts he defined with the terms *curiosité, expérience,* and *constant.* According to Graindorge, the scientific process began when an individual discovered an unexplained phenomenon in nature. He called such phe-

[47]Ibid., 639 (9 December 1665).

[48]Ibid., 551 (20 January 1666).

[49]For a variety of discussions illustrating the range of possible historical treatments of epistemological shifts affecting medical practitioners during this period, see Brockliss, *French Higher Education,* pp. 432–440; Claire Salomon-Bayet, *L'Institution de la science et l'expérience du vivant* (Paris, 1978); and Harold J. Cook, *The Decline of the Old Medical Regime in Stuart London* (Ithaca, N.Y., 1986). For an intriguing but controversial account of similar issues in experimental physics, see Shapin and Shaffer, *Leviathan and the Air-Pump.*

nomena *curiosités* (unverified claims or proposed "facts"). For Graindorge, these *curiosités* became research topics. Scientists needed to subject them to a series of *expériences* (trials designed to reveal inconsistencies or contradictions). When no *expérience* revealed inconsistencies, he accepted the *curiosité* as *un constant.*

The full development of this empiricism and the terminology that went with it was inextricably entwined with Graindorge's changing ideas about the dignities attached to various levels of scientific activity. As distinctions between practitioners dissolved in Graindorge's mind, the epistemology of *curiosités* became more highly articulated. When Graindorge began to see the common and mundane as the proper objects of scientific inquiry, the system built on *curiosités* became the means to establish priorities for a research program. In other words, the probability of finding *un constant* replaced the dignities of natural philosophy in determining the importance of research.

Besides replacing his hierarchical scheme for matching practitioners with research, the method required a new conception of the social organization of science. Since his entire system ultimately came to rest on the notion of the *curieux* as an individual participant in a Europe-wide scientific effort, the "academy" became his basic unit of organization. Indeed, working with *curiosités* could become feasible only within such a framework. Academies would ensure both that the necessary channels for scientific communication were available and that individual efforts did not dissipate into the mindless gathering of trivial facts.

Although Graindorge's empirical epistemology began to develop as soon as he arrived in Paris, his communal conception of science began to appear in an unmistakable form for the first time in November 1665—a full six months after he started to attend the Thévenot. Even then, Graindorge needed several weeks to arrive at the full articulation of his thoughts. The way he did so is noteworthy. Graindorge's letters show that he and Huet first became involved in a debate over the relative merits of *physique* versus *mathématique.* Their debate was one between an empiricist and a mathematical rationalist.

Huet had told Graindorge that he should learn more mathematics if he wanted to be a good natural philosopher. Graindorge allowed that this might be true but insisted on maintaining his own priorities. He would agree to study mathematics only if Huet would assure him that it was "useful to *physique,* because purely abstract speculations are meatless and hollow. I want us to put our hands to the work and visit the entrails of all sorts of animals. We have need of mutual encouragement to animate ourselves."[50] Graindorge emphasized the importance of group activity here, but significantly, the "mutual encouragement" he presented at this time was not yet the necessary basis for research—it served only as a stimulus for individual activity.

Through early November 1665, Huet and Graindorge continued their weekly exchanges on the relative merits of mathematics and *physique*.[51] Graindorge drew a sharp distinction between an individual mathematician reasoning on the orderliness of the cosmos (the task for *les mathématiciens*) and an organized, empirical probing of nature (the task of *les physiciens*). He clearly favored the latter. Huet responded with further arguments for the study of mathematics as the key to knowledge, and Graindorge countered with elaborations on his preference for empirical science:

> You are persuading me agreeably enough toward the study of mathematics, but I do not have a high opinion of this science as I do not like futile research. After a great deal of calculation and piling of figure upon figure, I do not see any fruit other than a chimerical proportion, which reveals nothing about the things that are real in nature. I am not at all satisfied to work on an idea made up of air to find nothing but air at the end. There are two issues: the one is the relation of mathematics to the real world, and the other is how it relates to pure thought. It is necessary to know the first in order to establish anything solid; the second creates only hollow truths.[52]

The following week Graindorge pushed this argument even further by placing his preference for empiricism squarely in the context of a concern for organized science for the first time. Nicholas Croixmare de Lasson, who would soon become one of the academicians in the Académie de Physique, had been in Paris during the entire time Graindorge was there. Although Lasson was a Caennais and an old friend of Huet's, it is clear that Graindorge had not known him very

[50] BL, A 1866, 644 (4 November 1665).

[51] Although there is no obvious way to establish any meaningful historical argument on the larger significance of this debate, a simple *post hoc, ergo propter hoc* suggests that these exchanges over the merits of *physique* versus *mathématique* mark the crucial turning point at which Graindorge persuaded Huet to support his program for the Académie de Physique. If that is the case, then this series of exchanges takes on larger significance in explaining Huet's commitment to the academy, his switch from astronomy to anatomy during 1665–1666, and ultimately his skeptical rejection of science as a key to metaphysics.

[52] BL, A 1866, 643 (11 November 1665).

well (if at all) before they met in Paris. When Graindorge wrote the following remarks, Lasson was preparing to leave Paris. Graindorge told Huet how much he would miss his new friend:

> I cannot praise M. de Lasson enough. . . . You know him better than I do, and I will only say to you that I could not value his candor and his talent any more than I do. It would be easy to bring about the return to the *siècle-d'or* if more could be found with his turn of mind. I have no greater desire than to be together with the two of you, since few things would escape our curiosity. It is certain that men mutually animate one another in the search for truth, and it is very difficult for one person working alone to succeed. Our start has been good enough; it will just require more of the same for us to make progress.[53]

From that point on, the themes of empiricism and the need for organized activity became touchstones for Graindorge's assessments of scientific activity.[54] He developed an almost missionary zeal in proselytizing his ideas, and when he returned to Caen in the spring of 1666 his enthusiasm for *curiosités* and organized activity set the program for the Académie de Physique.

In the scientific vocabulary Graindorge had developed while attending the Thévenot, a *curiosité* was a reported (but unverified) fact about nature. It was thus a research topic. Watching the scientific program at the Thévenot unfold over the course of a year was a startling experience for Graindorge. Not only did he add a great deal to his store of knowledge about nature but he also learned that the many "facts" he had taken for granted or assumed to be unassailable were simply erroneous opinions based on nothing more than folk wisdom, tradition, or superstition. He was excited by his new knowledge, but especially by any knowledge that allowed him to correct a false view of nature's workings. Hearing reports on Francesco Redi's research on snakes, for example, Graindorge was intrigued to learn that poisonous snakes are susceptible to their own venom when it enters the bloodstream and that humans can eat the meat of poisonous snakes without harm.[55] These *curiosités* told him that accepted ideas were wrong. For him, such claims offered almost endless possibilities for constructing new confirming *expériences*.

[53]Ibid., 642 (18 November 1665).

[54]For a comparison with England, see Marie Boas Hall, "Salomon's House Emergent: The Early Royal Society and Cooperative Research," in *The Analytic Spirit*, ed. Harry Woolf (Ithaca, N.Y., 1981).

[55]VDLI, pp. 308–312 (19 August 1665).

Exposure to the Thévenot led Graindorge to begin to doubt that traditional natural philosophy had been built on a firm foundation of facts about nature. It seemed to him that everywhere he looked, received knowledge about the world was in error. His only reasonable approach to the world in such a situation was through empiricism. At times he could carry this empiricism almost to a ludicrous extreme, as, for example, when he heard it was possible to drown a frog: "Someone said a frog will suffocate underwater if it stays there long enough. It is necessary to have the pleasure of attaching one to the bottom of a jar and filling the jar with water above the frog's nose—if frogs have noses—in order to see how long it can live before it suffocates."[56] As silly as the idea of such "research" may seem, the rationale behind it was quite serious. Nothing was so mundane that it could be taken for granted; nothing was so extraordinary that it could not be true.[57]

After his experiences at the Thévenot, Graindorge could not satisfy himself with any form of science that merely generated new knowledge. Rather, he insisted that any scientific inquiry must also offer the means to scrutinize the entire body of received knowledge. He wanted to use his science to create a better world—his *siècle-d'or*—but he was convinced that such a world would become possible only after the accepted wisdoms had been purged of fallacies. The doubts about "known facts" planted at the Thévenot thus led him to skepticism of a very particular type. He did not doubt that scientific truth existed, nor did he believe it was impossible to obtain; he just thought no one had yet arrived at anything close to it.

Once Graindorge convinced Huet of the validity of these ideas, the Académie de Physique had the basis for becoming a scientific re-

[56]Ibid., pp. 275–278 (3 June 1665).

[57]Examples of this attitude can be found throughout Graindorge's letters to Huet, but two examples in letters written in the early fall of 1665 are particularly graphic. In the first case, Graindorge explicitly asserted that studying marine anatomy seemed mundane and useless: "a quoy bon ces monstres et . . . tout ce grand attirail de poissons de mer qui vont et vienent et qui n'ont pas une once de sens commun." Then he immediately launched into a consideration of an important scientific question raised by thinking about fish: How can they fill their swim bladders with air when they live their whole lives underwater? When he opposed the notion that air might be "mixed" with seawater against the possibility that the fish transmute water to air, Graindorge found problems with both explanations. He left the issue unresolved but seemed to favor the "mixing" of air with water: "Je ne puis admettre la conversion dair en eau ny deau en air" (BL, A 1866, 1983 [23 September 1665]). Just weeks later he reported another curiosity—a plant that produced quicksilver—which seemed so fabulous it could hardly be believed. Nevertheless, he felt obliged to follow up by trying to obtain seeds from such plants. He doubted that the plants would produce mercury, but he thought "ce conte . . . meriteroit bien destre verifié" (ibid., 1982 [10 October 1665]).

search society. Given the support of Huet as patron, Graindorge's ideas about science gave the academy a significant research potential. When he returned to Caen in 1666, Graindorge would insist that the Académie de Physique existed to do new research, that the research had to be cooperative, and that each session must be devoted to research rather than "rare and curious" demonstrations or idle speculations. Such ideas were radical and would cause him problems, but before we deal with those problems it is important to conclude discussion of Graindorge's year in Paris by emphasizing its significance in a larger context.

THE SIGNIFICANCE OF PARIS

In Paris Graindorge became conversant with the latest scientific ideas from all over Europe, and his participation at the Thévenot brought the activities of Huet's group to the attention of scientists in the capital. Graindorge was virtually unknown in Paris in the spring of 1665, but by the fall of 1666, when Colbert was selecting the academicians for the new Académie Royale des Sciences, some of his friends in the capital nominated Graindorge for a chair.[58] The exchange between Paris and Caen was thus a two-way process. The Huet–Graindorge correspondence made the Académie de Physique a part of French science. Unfortunately, this situation did not last. Perhaps in this circumstance we may see why the Académie de Physique was the only royal academy of sciences created in the French provinces during the seventeenth century. Indeed, although the monarchy issued letters patent to several provincial academies of belles lettres over the course of the seventeenth century, it would not charter another provincial academy of sciences until 1706, when the Montpellier group became an adjunct to the Académie Royale.[59] It is on this issue that the real significance of Graindorge's experiences rests.

André Graindorge was not the only provincial *curieux* to attend the Thévenot during 1665–1666. His letters refer to intelligencers from both Dijon and Rouen who attended sessions. But he was among the last provincials given access to the new experimental science as prac-

[58]BL, A 1866, 665 [fall 1666].

[59]For a chronological table laying out the founding dates of provincial royal academies in France, see Daniel Roche, "Milieux académiques provinciaux et sociétés lumières," in *Livre et société dans la France du XVIIIe siècle,* ed. F. Furet (Paris, 1965), pp. 93–184.

ticed in Paris.[60] In late 1666, when Colbert organized the Académie Royale des Sciences, the open community of scientists that had been so important to Graindorge disappeared. Indeed, Thévenot closed his academy in March 1666, as soon as he learned that Colbert had definitely committed the monarchy to a royal institution.[61] In the process of establishing the Académie Royale, Colbert selected just fifteen academicians, and when the institution opened its sessions, it exercised a virtual monopoly over the practice of science in the capital. All deliberations were held in secret; all publications appeared anonymously; and outsiders were allowed into the sessions only upon special invitation to have their work evaluated. Visitors could present their finished research but could not hear the deliberations on its merits.

Even before the academy officially opened its register on 22 December 1666, the Parisian community of scientists had changed in response to expectations about the new royal institution. Besides the closing of the Thévenot in March, news of the royal institution had other effects on the community of Parisian amateurs. Adrien Auzout told Henry Oldenburg, for example, that many scientists in Paris had held off on new research during the summer of 1666 in anticipation of performing their work as royal academicians.[62] As early as the spring of 1666 Paris was full of rumors about Colbert's plans for "la Grande Académie."[63] Everyone expected great things from it, everyone hoped to be part of it, but no one seemed to have much information about who would participate. As late as December 1666, at the time of the first official sessions, the only thing Adrien Auzout could really tell Oldenburg was that he (Auzout) had been selected for membership. Even as an academician, he still had no idea how the academy would function.[64]

[60]Through his frequent references to Jean-Baptiste Lantin, his interest in optics during the summer of 1665, and his references to M. Avoye in Rouen (winter 1665), Graindorge reveals that the Thévenot maintained relations with amateur groups in Dijon and Rouen as well as with the group in Caen. This is a subject that warrants further research.

[61]BL, A 1866, 650, 662, 649 (17, 22, and 24 March 1666). In these three letters Graindorge chronicled the end of the Thévenot. The first describes that academy's last session. The second announces Colbert's plan to open the "Grande Académie." The third reports that Thévenot has stopped hosting his group.

[62]Auzout to Oldenburg, 18 December 1666, *CHO,* 3:291–295, 589.

[63]This is the term by which Graindorge identified the project that developed into the Académie Royale des Sciences at the end of 1666.

[64]Auzout to Oldenburg, 18 December 1666, *CHO,* 3:291–295, 589.

The primary reason for the secrecy surrounding the planning for "la Grande Académie" seems to have been the fact that Colbert wanted to found more than simply an academy of sciences. He planned to open a universal academy incorporating four sections: philosophy, literature, history, and mathematics. His plans met with opposition from established corporations such as the university, the parlement of Paris, and the guilds.[65] In the end, Louis XIV would approve only those parts of the plan that called for an academy of sciences. As soon as the academy opened with its first session on 22 December 1666, the entire community of European scientists began to speculate about what was going on behind the closed doors of the king's library. Everyone expected marvels, but no one had much information other than the little the academy chose to announce.[66]

As difficult as it may be to believe, the cloak of secrecy surrounding the scientific program of the Académie Royale was effective. Henry Oldenburg was eager to get any news he could about the new academy in Paris, but a full year after it had opened he still had no clear idea of how it had been structured or who its members were. In Caen they knew almost nothing. There was a great deal of speculation about the activities at the new academy, but no one outside the academy itself knew very much about its projects. In André Graindorge's words: "Ils font grand mystère."[67]

Given the importance of Graindorge's experiences in Paris for the Académie de Physique, the incorporation of the Académie Royale des Sciences must be interpreted as a significant impediment to the flow of scientific expertise from the capital to the provinces. There is no question about the importance of Graindorge's experiences for the Académie de Physique; yet shortly after he actually went through those experiences, it became virtually impossible for him or for any other provincial *curieux* to do the same. Perhaps the best evidence for that fact comes from Graindorge himself. During his 1665–1666 stay in Paris he told Huet that he did not mind being away from Caen for so long because "if I bring home no other fruit than the passion for natural philosophy, which redoubles itself in me every day, I shall not complain of my trip."[68] Less than two years later (but after the open-

[65]George, "Genesis," pp. 379–386.
[66]Hahn, *Anatomy of a Scientific Institution,* pp. 16–17.
[67]BL, A 1866, 572 (16 January 1668).
[68]VDLI, pp. 305–308 (13 August 1665).

ing of the Académie Royale) he returned to Paris but wrote to Huet about Parisian science in terms that stand in stark contrast to his earlier statement: "Letters are in a profound silence on whatever side I turn. I am learning nothing new, and I would rather dissect a lobster or a shad with the rest of you than run about the streets here finding nothing."[69]

The incorporation of the Académie Royale des Sciences destroyed an open scientific community that welcomed such provincials as Graindorge. His own letters amply document that fact. Its opening meant the end to a group such as the Thévenot. For Graindorge these events were clearly linked. He told Huet about the coming of the Académie Royale and the closing of the Thévenot in the same dramatic announcement: "The Grand Academy will be formed; M. Thévenot is doing nothing."[70] This tantalizing statement indicates that Graindorge drew the connection between the coming of the Académie Royale (still "la Grande Académie" in March 1666) and the closing of the Thévenot, although it does nothing to explain why he associated these events. Such an omission seems unfortunate for the modern historian who seeks to deal with the Fontenelle thesis and its assertion that the collapse of French patronage led to the formation of the royal institution.

Of course, Graindorge labored under no obligation to explain his thoughts for the benefit of posterity. His only obligation was to his patron, Huet. Yet in that sense his silence on the imperative mandating an end to the Thévenot actually becomes evidence that suggests his reasoning. Because the total lack of analytical commentary was one of the most striking features of all Graindorge's discussions on the planning for "la Grande Académie," Huet must also have known exactly what Colbert's plan implied. Understanding appears to have passed tacitly between Graindorge and Huet, and if it did, there is only one really cogent explanation: Both Graindorge and Huet had attuned themselves to the new royal policies toward patronage.

Louis XIV's determination to serve as his own first minister following Mazarin's death in 1661 marked a genuine turning point in the political history of France.[71] It was an announcement that shocked

[69]BL, A 1866, 682 ("dernier avril" [1667]).

[70]Ibid., 662 (22 March 1666): "La grande Academie fera M. Thevenot ne fait rien et peu de chose se fait shez les particulieres."

[71]For discussion of the political transformations that affected France during the 1660s, see John B. Wolf, *Louis XIV* (New York, 1968), pp. 133–181. For a historiographic treatment of Louis XIV's reign, see William F. Church, *Louis XIV in Historical Thought* (New York, 1976).

the highest circles of the feudal nobility and the court. Following up on that decision barely six months later, the young king again electrified all of Europe when he arranged for D'Artagnan and his musketeers to make the dramatic arrest of the powerful superintendant of finance, Nicolas Fouquet, while also ordering royal troops to seize the minister's fortified château at Belle-Isle.[72] Such an action sent shock waves rolling through the very highest levels, and aftershocks reverberated almost instantly through the Republic of Letters and the world of patronage. The day after Fouquet's arrest, for example, the artist Charles Le Brun signaled acquiescence to his patron's fall as he begged Madame Colbert to accept the gift of a drawing.[73]

Louis XIV's determination to rule in his own right was a political decision, but it was one that had far-reaching effects throughout French society. Vast patronage networks had formed a natural part of a political system in which great ministers such as Richelieu, Mazarin, or Fouquet and peers such as the Great Condé and the duc de Longueville could jockey for power, or even take up arms against royal troops while claiming to support the monarchy's true cause—as had happened in the tragicomic Princes' Fronde. Louis XIV's decision to act as his own first minister brought that political system to an end. Moreover, with Fouquet's fall, anyone powerful enough to establish an extensive retinue of clients had been put on warning; after all, as Fouquet's trial demonstrated, his one unforgivable crime against the king had been simply to display too much wealth, too much ostentation, in his building program and his patronage.[74]

The early 1660s were years when all Europe watched Louis XIV restructure the French Republic of Letters. Beginning in 1661, Colbert's rise, the annual *gratifications* to savants, the formation of the Academy of Inscriptions, the new letters patent for the Académie Royale de Peinture et de Sculpture,[75] the plan for "la Grande Académie"—all signaled the formulation of a clear-cut royal policy toward the arts and the Republic of Letters. Just as the king's determination to act as his own first minister dictated the elimination of all political clientages other than his own, the same decision rippled down through the Republic of Letters. The king's patronage would be the greatest; there would be no meaningful competition.

[72]Wolf, *Louis XIV*, pp. 137–142.

[73]David Maland, *Culture and Society in Seventeenth-Century France* (London, 1970), p. 175.

[74]See Paul Morand, *Fouquet ou Le Soleil offusqué* (Paris, 1961).

[75]See Murat, *Colbert;* Meyer, *Colbert;* Maland, *Culture and Society*, pp. 233–240.

Thus as Colbert revealed his plan for "la Grande Académie," there was every reason for Thévenot to close his academy. If Graindorge failed to explain those reasons to Huet, it was because there was no need. Thévenot was himself a prime candidate for membership in "la Grande Académie," especially given his linguistic skills. To have continued to host his academy would have been to go against the grain of Colbert's plans. Anyone with the vaguest sense of how the political winds blew in the early 1660s would have been loath to do such a thing, particularly if he thought he might have a chance at gaining a chair in the new royal organization.

Colbert's plan for "la Grande Académie" ultimately proved to be the issue on which the corporate interests in Paris blunted the royal initiative in the Republic of Letters.[76] Nevertheless, the new Académie Royale that emerged from the remains of that ambitious project started its work with virtually all private-patronage competition wiped from the capital.[77] Moreover, with its hand-picked membership, its closed sessions, and its anonymous publication, the Académie Royale's own organizational form militated against the reemergence of meaningful private-patronage groups. Who could compete with such resources? How could anyone expect this organization to share its research results?

The scientific situation in Paris had changed and the consequences of this change were far-reaching, both in the internal development of French scientific thought and in the social organization of French science. Within two years, for example, many of those whom Graindorge mentioned as attending the Thévenot during 1665–1666 were no longer active in research or (like Adrien Auzout and Thévenot himself) no longer lived in Paris. By the time Graindorge left Paris, the closing of the Thévenot had made impossible the kind of transfer of basic knowledge and techniques which had occurred so easily through the Huet–Graindorge correspondence.[78]

[76]George, "Genesis," pp. 386–392; Hahn, *Anatomy of an Institution,* pp. 11–15.

[77]Brown, *Scientific Organizations,* pp. 161–184, 231–253.

[78]BL, A 1866, 650, 662, and 649 (17, 22, and 24 March 1666). Graindorge's comments in these three letters show that the last session of the Thévenot was held on 16 March 1666. At this last session, the members of the Thévenot investigated the hygroscopic properties of salts (letter of 17 March 1666).

3

The Dynamics of a Scientific Organization

André Graindorge expected to see Huet when he returned to Caen in mid-April 1666. Unfortunately, Huet had left the city early on the very morning Graindorge arrived.[1] At that point Huet and Graindorge had not seen each other for almost a year, and although Graindorge did not know it then, another nine months would pass before they met again.[2] That fact proved to have important consequences for the Académie de Physique. Although Huet had built the "academy" around the scientific program Graindorge's letters had provided, Graindorge had never actually been present at a session at which his program was followed, and he had no experience at putting his grand ideas into practice. When Huet left him to direct the sessions, the task proved much more difficult than he had envisioned. Huet's absence put Graindorge in a difficult position. He had to make the academy's program work, but he did not really know how.

In Paris, Graindorge had been a neophyte provincial who merely followed the program laid down by his more experienced mentors at the Thévenot. He had watched Niels Steno dissect, for example, but he had never actually had the chance to take a hand in the cutting necessary to reproduce the Dane's marvelous anatomical demonstrations. Likewise, he had peered through microscopes and telescopes

[1]BL, A 1866, 541, 548 (10 and 19 April 1666).

[2]Altogether, between Graindorge's departure for Paris (early May 1665) and Huet's return to Caen (mid-January 1667), they passed twenty-one months without meeting face to face.

with Adrien Auzout and seen numerous chemical and physical demonstrations, but he had never set up the instruments or equipment for any of them. Moreover, Graindorge had borne no responsibility for setting the Thévenot's scientific agenda or organizing its sessions. He had merely arrived at those sessions and joined in the activities. All that changed when he returned to Caen. Huet had left him a small but thriving scientific academy. Clearly he was in for a difficult time.

The host of problems that Graindorge would face before Huet returned were of two sorts: first, he had to galvanize the academy's membership to work under his direction rather than Huet's, and second, he had to translate his knowledge of Parisian science into a sustained program of new scientific research. To deal with these problems Graindorge was armed only with his Paris experiences and the modest authority provided by his role as Huet's *fidèle*. Very quickly he discovered that these assets were inadequate for the task at hand. In his definition of an academy, the purpose of sessions was research. The members should come prepared to set up *expériences* and to learn something new from each of them. For Graindorge, marvelous demonstrations and public spectacles were useful only if they led the academicians to more substantial work. In trying to "animate" Huet's academicians with such a purpose, he faced his first challenges as secretary of the Académie de Physique.

THE ACADEMICIANS

At the time Graindorge took over responsibility for directing the sessions of the Académie de Physique, that organization boasted a membership of at least seven.[3] besides Huet and Graindorge, four

[3]No formal list of the academicians who attended sessions exists for the period before royal incorporation (1667). Those included in this discussion are mentioned in Graindorge's letters between May 1665 and January 1667. It is possible that others attended sessions when Huet was in Caen but refused to participate under Graindorge's leadership. Circumstantial evidence (Villons's behavior and the similar behavior of others after the royal incorporation) makes it seem likely that Huet's sessions had attracted more participants. Slightly better evidence is supplied by the fact that during early 1666 Graindorge's letters from Paris mention other names in association with the scientific activity at Huet's house. His letter of 3 February 1666, for example, in which Graindorge queried the opinion of Corneille Vicquemant, a *médecin,* on a matter concerning vision, suggests that Vicquemant was attending Huet's sessions at the time (BL, A 1866, 553). The five participants discussed in this section (four academicians and Busnel) were all men named by Graindorge at some point in 1666 as having attended a session. During 1667 the membership would grow to ten.

other men can be called its regular academicians. In addition, Huet claimed in his *Commentarius* that since 1662 he had employed the services of a surgeon who acted as the group's anatomical demonstrator. The surgeon, Charles Busnel, was still active in early 1666, but it appears Graindorge's new ideas on organizing research—he intended that he and Huet should personally "visit the entrails" of various animals—had seriously limited Busnel's role in the academy. Although Graindorge's letters give little specific information on how these men perceived their academy, it is clear that they operated without any formal rules or written bylaws. They did not need either. The group had simply coalesced around the scientific program Huet had taken from Graindorge's reports on the Thévenot. All those involved had reason to defer to Huet and to give him the respect due someone of his stature in the social and intellectual hierarchies of Caen. His position as the organization's patron provided the only authority he needed to guide the academicians who attended his sessions. With Graindorge acting as *chef,* however, the situation changed. Several of those who eagerly participated in Huet's sessions must have felt uneasy as they pondered the prospect of working under Graindorge's direction. He was not their patron, and he did not possess the *état* to become their *chef.*

One man who never submitted willingly to Graindorge's authority was Jean Gosselin, le chevalier de Villons.[4] Although the chevalier was clearly an academician in 1666, his name does not appear in any of Graindorge's accounts of sessions for several months after Huet's departure. Nor did this boycott during Graindorge's first term as the academy's acting *chef* prove an aberration in Villons's behavior. In every succeeding period when Graindorge took over control of sessions, Villons followed the same course—he simply stopped attending the academy's sessions. Frequently he refused to communicate with Graindorge, or even to let him know what kind of projects he was working on.[5]

The loss of Villons's participation during those periods when Graindorge was directing the sessions was a serious blow to the Aca-

[4]For biographical information on Villons (1619–?), see Tolmer, *Huet,* p. 355.

[5]Villons remained aloof from the academy whenever Graindorge was in charge until, without notifiying Graindorge, he resigned his chair in 1670. Graindorge learned of Villons's resignation at secondhand (BL, A 1866, 702 [late 1670]). Graindorge's report on what Guy Chamillart told him about Villons's resignation at that point summarizes the man's attitude toward Graindorge: "Il ma dit quil ne prend pas plaisir a dire les chose quil fait mais quil a bien parté et de notre Académie en general et de moy en particulier."

démie de Physique. By common assent, Villons was a skilled mechanician—he could fabricate precision instruments and build intricate mechanical devices. To borrow one of Graindorge's own phrases for describing someone with similar skills, he possessed "prodigious ingenuity in the tips of his fingers."[6] Less colorfully, Villons was described for the king's minister Colbert as "a genius with machines."[7] Such a man could have proven extremely useful in helping Graindorge fulfill his plans for an empirical, instrument-based scientific program. Villons worked on a number of important projects, and in fact in 1670 his success in designing a new drive mechanism for clocks would lead to a royal appointment and a commission to build one of his clocks for Louis XIV.[8] The chevalier de Villons was active when the academy was under the direction of its patron; he simply would not give his support to Graindorge and his projects.[9]

Villons was the extreme case in his refusal to work with Graindorge, but another academician who also showed reluctance to become too involved with the academy during Huet's absences was Nicholas Croixmare, sieur de Lasson.[10] Ironically, it seems that Graindorge was responsible for recruiting Lasson as an academician. He had been in Paris when Graindorge arrived there in May 1665, and the two men had frequented the various scientific circles in Paris together until Lasson returned to Caen in November 1665. When Graindorge returned home the following April, Lasson was definitely among Huet's academicians.[11] Graindorge counted on his attendance. Clearly Lasson and Graindorge were friends on a personal level. Nevertheless, Lasson was less than enthusiastic about serving as an academician under Graindorge's direction. His attendance at Graindorge's sessions was sporadic at best, and in early 1668 Graindorge's offer to

[6]BL, A 1866, 568 (12 December 1667). Graindorge would use this phrase to describe another academician, Jean-Baptiste Callard, sieur de La Ducquerie.

[7]Guy Chamillart to Colbert, 28 February 1669, BN, Mélanges de Colbert, 149 bis, f. 661.

[8]BL, A 1866, 655 [late 1670].

[9]For example, as soon as the royal intendant Chamillart agreed to become the new *chef* of the academy (November 1667), Villons contacted him and told him about his work. All Graindorge knew was that "M. l'Intendant me dist bien que le chevalier travailloit sans nous vouloir decouvrir sur quoy" (ibid., 566 [23 November 1667]).

[10]For biographical information on Lasson (1629–1680), see Georges Huard, *Deux Académiciens caennais des XVIIe et XVIIIe siècles: Les Croismare, seigneurs de Lasson* (Caen, 1921).

[11]BL, A 1866, 642, 549 (18 November 1665, 5 January 1666). In his letter of 5 January Graindorge sent greetings to Lasson through Huet. From that point until his return in April, Graindorge's letters make several references to Lasson.

underwrite the cost of supplies he needed in order to cast a burning mirror insulted him so much that he stopped attending altogether.[12]

Like Villons, Lasson was a mechanician. Also like Villons, he would be recommended to Colbert as a man who could produce marvelous mechanical devices. In his case, he was expected to fabricate telescopes, a "burning mirror" larger than any yet cast, and precision astronomical instruments.[13] Lasson, however, had a reputation as something of a gadfly and scientific prankster. He was far more successful with these endeavors than with anything he did for the academy. For example, Lasson produced a hoax on the comet craze which was sophisticated enough to dupe Christiaan Huygens. Moreover, he probably authored another similar hoax, a report on stone-eating worms, which duped the editors of the *Journal des Savants.*[14] Some also suspected him of authoring a pair of scurrilous satires lampooning Jean Chapelain.[15] In all probability, he did not write these two works, but his reputation was such that his contemporaries believed he did. Lasson was considered brilliant, if somewhat erratic, an opinion Pierre-Daniel Huet gave polite expression by saying, "His spirit partook of so many occupations and knowledges that it skimmed the surface of everything and plumbed the depths of nothing."[16] Lasson's talents offered the Académie de Physique a great deal; with Graindorge as *chef,* he produced nothing. It is impossible to say how much he might have accomplished had Huet continued to direct the academy's sessions.

Both Villons and Lasson had reason to look on Graindorge as a parvenu overstepping himself when he presumed to substitute for Huet, but they also had another reason for avoiding the academy when he was in charge. Both men shared Huet's enthusiasm for mathematical rationalism as the soundest approach to nature.[17] Thus

[12]Ibid., 642, 571, 573 (18 November 1665, 13 and 27 January 1668). From this point on, Lasson boycotted the academy's program whenever it was under Graindorge's direction.

[13]Ibid., 670 [1671].

[14]Tolmer, *Huet,* pp. 301–304, 312. The account that appeared in the *Journal des Savants,* no. 32 (9 August 1666), was a spoof on the recent enthusiasm for microscopy and reported such astounding observations as that microscopic investigation revealed "crumbs" left by worms who had eaten a stone sealed in a box. The stone had not been totally consumed, however; it was only "sensibly smaller" than it had been at the time it was put in the box.

[15]*Chapelain décoeffé* and *La Métamorphose de la perruque de Chapelain en comète* (Tolmer, *Huet,* pp. 301–304).

[16]Huet, *Origines,* p. 429.

[17]Their presence at Huet's sessions would lend some credence to his claim in the *Origines de la ville de Caen* about the importance of the comets for the gathering of

three of the academy's six members were *mathématiciens*. Although Graindorge obviously had to grant legitimacy to *les mathématiciens* in order to deal with his patron, Huet, the debate over the relative merits of *mathématique* versus his own *physique* was one in which he was always willing to engage—even with Huet. On a more practical level, Graindorge's vehemence on this subject would have offered grounds for Villons and Lasson to boycott sessions even if they had had no other reason for doing so. With Graindorge setting the academy's agenda, little time was given to the "hollow truths" of mathematics.[18] Nevertheless, neither seemed willing to press that point with someone of Graindorge's low stature. It appears they found it easier simply to stay away from the academy when he was in charge. With Huet gone much of the time, the academy thus had no *mathématiciens* under Graindorge, and at best functioned at half strength.

Graindorge had fewer problems in dealing with the three other men who can be counted as members of the Académie de Physique. Along with Graindorge, these men constituted the section for *physique*. Two of them, Pierre Hauton and Matthieu Maheust de Vaucouleurs, took their responsibilities as academicians seriously. For long periods at a time, both attended regularly; both contributed to the success of Graindorge's program. Unfortunately, Hauton and Vaucouleurs detested each other and frequently became involved in bitter quarrels outside the academy.[19] It is unclear whether their animosity had its roots in personal differences, but certainly their feud came to a head in professional disputes. Both men were *médecins*, but with that general professional title all similarities between them ended. Hauton, an iatrochemical practitioner, was given to "the search for the philosopher's stone."[20] Graindorge's letters confirm that assessment, at least for the period around 1665.[21] Vaucouleurs

---

group activity. No evidence documenting any sustained program of astronomical observations has survived, however.

[18]In 1670 Graindorge's unwillingness to use the academy's money to purchase astronomical instruments would divide the academy. See chap. 6 below.

[19]It seems both were active with the early *assemblée*. Both are mentioned in Graindorge's letters from Paris, but more significant, they can be tied to Graindorge and Huet as early as 1661 (VDLI, pp. 255–266 [11 and 24 April 1661]). The first mention of their hostility appears during the summer of 1665 (ibid., pp. 278–281 [10 June 1665]). Graindorge says that at that point Vaucouleurs was pressing a lawsuit against Hauton, and he indicates that both he and Huet had tried to dissuade Vaucouleurs from pursuing the case.

[20]Huet, *Commentarius*, pp. 223–227.

[21]See, for example, Graindorge's letter of 3 June 1665 (VDLI, pp. 275–278), in which he concludes a discussion of a chemical *expérience* he has heard about by sug-

was not only a traditional Galenist but also a professor of medicine at the university in Caen.[22] As partisans in the debates over iatrochemical medicine, these men had every reason to be at each other's throat.[23]

Given the hostility between Hauton and Vaucouleurs, their dealings with each other at the Académie de Physique were very courteous for a very long time. The civility of their relationship can be attributed in part to Graindorge. Hauton was not active during the second half of 1666,[24] but later both he and Vaucouleurs proved willing to work under Graindorge's direction, and it seems that they worked in close proximity without any serious altercations for several years because they were committed to the success of his scientific program. The alchemist Hauton, for example, gave up his search for the philosopher's stone for the two years it took to perfect an apparatus for desalinating seawater. The idea for this project originated with Graindorge, and Hauton proved willing to complete it while Graindorge urged persistence in the face of various disappointments.[25] His efforts would earn praise from the Académie Royale des Sciences and a substantial *gratification* in the name of Louis XIV. Vaucouleurs, for his part, would complete a number of projects, and aside from Graindorge, he was the academician most consistently active in the group's anatomy program.

---

gesting that Hauton might try to make it work: "Quand même il ne s'y formerait pas des pierres précieuses ni de ce riche métal, . . . nous nous contenterions du commun."

[22]For biographical information on Vaucouleurs (1630–1700), see Tolmer, *Huet*, pp. 351–355.

[23]For discussion of the iatrochemical debates and their impact on the Galenist position across the seventeenth century, see Brockliss, *French Higher Education*, pp. 391–440; François Duchesneau, "Malpighi, Descartes and the Epistemological Problems of Iatromechanism," in *Reason, Experiment, and Mysticism in the Scientific Revolution*, ed. M. L. Righini Bonelli and William R. Shea (New York, 1975). For an introduction to medical practices during this period, see Marcel Sendrail, "La Médecine au grand siècle," *XVIIe Siècle*, no. 35 (1957), pp. 163–170.

[24]Graindorge's letters from the period before his return to Caen do indicate Hauton was attending Huet's sessions. See, for example, his letter of 3 February 1666 (BL, A 1866, 553), in which he described a chemical *expérience* he thought Hauton might want to try.

[25]BL, A 1866, 562 (4 July 1667). Originally Graindorge suggested that Hauton turn his specialized skills to the distillation of alcohol, but by November 1667 Hauton was working on the distillation of seawater (ibid., 564 [11 November 1667]). The water Hauton produced was potable, but he had a difficult time improving its foul taste. Both Huet and Graindorge became anxious about the project because, while Hauton was working, several other *curieux* took various devices to the Académie Royale for trials in the hope of claiming the prize offered for a desalinization process that could provide safe drinking water for ships at sea. See chap. 7 below.

Ultimately, however, the professional debate between Hauton and Vaucouleurs would get the better of everyone. During an epidemic in the early spring of 1670, they became involved in a dispute over the value of venesection after one of Hauton's patients died without being bled. At that point, their long-running debate finally spilled over into a shouting match at one of the academy's sessions. When Graindorge intervened and told them they were opening the academy to ridicule, they were both offended; both stopped attending sessions.[26] Apparently Graindorge's personal authority was limited even with those who were inclined to follow his lead in the scientific program.

Graindorge faced a very different sort of problem in dealing with the only other man who can be described as active in the Académie de Physique as early as 1666. This was Charles Busnel, the surgeon who served as the academy's anatomical demonstrator.[27] He was clearly not an academician. In the kind of academy Graindorge envisioned, Busnel held a distinctly inferior role—if there was any need for him at all. Huet's position as patron in the early *assemblée* had called for him to retain the services of someone like Busnel to handle all but the most delicate and important parts of his dissections.[28] Watching Niels Steno dissect at the Thévenot during 1665, however, had given Graindorge a new perspective on the usefulness of someone like Busnel. When Graindorge claimed he wanted "to visit the entrails of all sorts of animals" with Huet, he meant the statement literally.

The active involvement of the scientist in every detail of his own research was central to the way Graindorge had conceived the difference between his new form of empirical science and traditional natural philosophy. Thus Graindorge's notion of what the academy should become put Busnel in a very difficult position. He was Huet's

[26]BL, A 1866, 599 (1 April 1670).

[27]Huet claimed (*Commentarius,* pp. 220–221) that from 1662 he had employed a demonstrator who had access to human cadavers by virtue of his position as a surgeon at a hospital. In 1665 letters Graindorge wrote from Paris describe his inquiries about the cost of buying a *maîtrise* for the surgeon Charles Busnel, who was serving as Huet's demonstrator at that point (VDLI, pp. 275–278, 319–322 [3 June and 16 September 1665]). It seems reasonable to conclude that Busnel had been dissecting for Huet since the earliest days of the *assemblée.* Virtually no biographical information on Busnel has survived, except that he was a surgeon *juré,* who was granted a *survivance* as *chirugien de l'Hôtel Dieu* in 1671. He held that post until 1691 (G. Dupont, *Registres de l'Hôtel de ville de Caen: Inventaire Sommaire,* vol. 5 [1665–1724]). Graindorge's letters indicate that Huet helped Busnel obtain the *maîtrise* in his guild.

[28]Huet himself makes this claim in the *Commentarius,* p. 220.

demonstrator and was probably paid for his professional services at the sessions, but under Graindorge he had no place. With Graindorge setting the agenda, Busnel faced an unpleasant choice. To attend, he had either to accept a demeaning and unprofessional role as a mere assistant or to "put on airs" by acting more like a regular academician. Neither alternative would have been attractive to the surgeon. Of course, he too could exercise the option of boycotting sessions directed by Graindorge. It appears he frequently chose that course.

Thus as Graindorge faced the prospect of taking control of Huet's academy in April 1666, he must have done so with trepidation. Of the four academicians it is possible to identify, at least two considered him a social inferior and opposed his ideas on the scientific program, and the two who were willing to work with him were hostile to each other. With Busnel, Graindorge wanted to redefine the role the surgeon was accustomed to performing in the academy. Indeed, Graindorge did have reservations about his ability to keep the group functioning during Huet's absence. A month before he left Paris he told Huet about his fears, and in his last letter before coming home Graindorge admitted that he knew the academicians were not "very warm" about the prospect of his taking over the academy.[29]

The idea of trying to act as *chef* to the Académie de Physique intimidated Graindorge. Once back in Caen he dropped the impressive "académie" when he referred to the group, for a time calling it simply "notre assemblée physique" instead. Graindorge had good reason for beginning to doubt that there actually was an "academy" in Caen. He was unable to organize a session for the first Thursday after he returned, and when he tried for the next week, Lasson claimed he was incapacitated by gout, and both Hauton and Vaucouleurs were unavailable.[30] He made no mention of even trying to contact Villons.

Graindorge finally managed to organize his first session on 29 April 1666. Apparently the fact that two *curieux* he had met in Paris had come to visit him in Caen aided his efforts considerably.[31] Undoubt-

[29]BL, A 1866, 541, 648 (10 April and 10 March, 1666).

[30]Ibid., 548, 555 (19 April and 7 May 1666).

[31]One of the visitors, Martin Fogel (1634–1675), was a *médecin* whom Graindorge described for Huet as "un tres galand homme de Hambourg. . . . Il s'apelle Fogle ou Vogel. Il a une maitre lespagne et la france avec un esprit dallemant et est remply de touttes les connoissances qui regardent les bonnes lettres aussy bien que les belles. Il a bien de candeur et il na pas besoin de recommendation pour estre bien receu de vous. Il est philosophe, médecin, chyurgien, mathematicien, curieux de livres et de tout ce que lon peut savoir" (ibid., 541 [10 April 1666]). The other visitor was a M. Avoye from

edly the presence of these visitors, who were actually part of a larger party of Germans touring France, provided an attraction that allowed Graindorge to organize the session fairly easily. In any event, he did manage to direct a session. It was not entirely successful, but by accomplishing as much as he did with it, Graindorge overcame an important hurdle. He had brought Huet's academicians—or at least most of them—together at a session under his direction. He would have a good many problems with the academicians in the next few years, but he must have felt very good just to have gotten through that first session.

GRAINDORGE'S SCIENCE

Even as Graindorge revealed doubts about the "warmth" the academicians would show him as he tried to become their *chef,* he expressed confidence in the scientific program he offered them. He claimed he knew about a "good many curiosités."[32] He was familiar with all the current research topics under discussion in Paris, and through his participation at the Thévenot he had gained as much familiarity with the current work of both the Cimento and the Royal Society as anyone in France could claim.[33] In Graindorge's opinion, this knowledge from the Thévenot was a valuable commodity. He planned to use it to good advantage.

At the level of day-to-day practice, the empiricism Graindorge had developed in Paris called for trials or *expériences* that could confirm *curiosités.* During his time at the Thévenot he had heard a great many claims about various natural phenomena. His return to Caen gave him the opportunity to probe some of these claims. He had been particularly impressed by the reports he had heard and read on Francesco Redi's work with snakes and poisons.[34] For his first session as the academy's acting director, he scheduled an *expérience* that he

Rouen, whom Graindorge had identified for Huet as a participant at the Thévenot in December 1665 (ibid., 547 [16 December 1665]).

[32]Ibid., 541 (10 April 1666).

[33]For information on the scientific communications maintained by the Thévenot, see Brown, *Scientific Organizations,* pp. 135–136. Although Brown's discussion is limited to the period before Graindorge began to attend the Thévenot, Graindorge's letters clearly show that the correspondence between Paris and Florence was maintained throughout 1665 and into early 1666.

[34]VDLI, pp. 308–312 (19 August 1665). Graindorge devoted this entire letter to Redi's work after introducing the subject by saying: "Je viens de lire les observations d'un Italien sur les vipères. Il me semble que je ne vois autre chose. Je suis tellement

thought would test some of Redi's claims about the poisonous effects of an extract of tobacco.[35] According to Graindorge's understanding, ingestion of this oil of tobacco should at worst make a small animal only mildly ill, but should be immediately fatal if it entered the circulatory system directly. Thus the oil of tobacco should act on a pigeon in the same way Redi claimed poisonous venom acted on a snake. He secured a specimen and prepared his extract in advance, but the trial did not work entirely as he had expected:

> We put the oil on the tongue of a pigeon, which fell stunned and made a thousand grimaces. At first it had convulsions fairly frequently, but it revived little by little and finally voided heavily. After which, we pricked the pigeon under the wing and applied some of the oil to the wound. This made the pigeon recommence its senselessness, but without dying.[36]

Despite what might seem a rather disappointing result—the pigeon became very ill when Graindorge gave it the poison orally, but did not die when he applied it to the wound—Graindorge was pleased by the session. He could think of only two explanations for his failure with the oil of tobacco when it had killed animals in both Florence and London: "It must be either that their oil of tobacco was of a finer quality, or that the pigeon was immune to the effects." Having arrived at that conclusion, Graindorge then claimed that his failure with the *expérience* confirmed just how important his approach to science was: "Thus, this shows it is necessary not to rush too quickly to believe all those things someone has sworn to be constant."[37] He had struck the academy's first blow at false and erroneous scientific knowledge.

plein que je ne vous entretiendrai d'autre chose, car que peut sortir d'un sac que ce qu'il y a? Vous savez ce que les auteurs disent de ces animaux, et qu'ils en content mille choses qui passent de main en main pour des vérités insoutenables, et cependant, comme l'on n'en vient point à l'expérience, l'on croupit dans des erreurs insupportables." Clearly Graindorge was impressed by this work. He referred to it numerous times before returning to Caen the next spring.

[35]His letters in early February raise the question of what effects poisons have on their victims' blood. In his letter of 17 February he summarized his understanding of the effects the oil of tobacco should have: "Il s'en faut tenir a ce que Je vous en mandé pour lors verifieray dans Redi. L'on dit que les poules et plusieures animaux piques jusques au sang meurent incontiment quon leur applique sur la playe, qu un filet trempé dans cette huile et passé avec une aiguille dans la cuisse dune poule la tue sur le champ. Un chien baté en sera beaucoup plux joyeux" (BL, A 1866, *inserto* 651).

[36]Ibid., 555 (7 May 1666).

[37]Ibid. Besides attempting the *expérience* with the oil of tobacco at Graindorge's first session, they had also dissected a dog in an attempt to trace parts of its lymphatic system. Graindorge claimed the dog was too small to permit them to see anything very well.

Graindorge was pleased when his demonstration contradicted the reports from Florence and London, but he probably had difficulty convincing the others who had gathered to witness this *expérience* that he had achieved a desirable outcome. Undoubtedly most would have preferred to see the results that had been promised by the reports from the Cimento and the new Royal Society in London. At the time Graindorge returned from Paris, it was possible for a "rare and curious" scientific demonstration to attract a considerable audience, most of whom would have come for the promise of a spectacle. In early May, for example, Graindorge reported to Huet that the academy had missed the opportunity to do a wonderful dissection when he (Graindorge) had failed to gain control of a specimen quickly enough. Some local fishermen had brought in a very large fox shark, which they offered for sale. Samuel Bochart had bought it and arranged a public dissection to be done by a local surgeon at the house of a certain Madame Luzerne. Graindorge claimed that the sight of this shark "made his mouth water," but that he had learned nothing from watching the dissection because the surgeon, while skillful at cutting, was extremely ignorant of anatomy.[38]

Despite his feelings about research, Graindorge was not too proud to engage in such theatrics. Indeed, everyone expected him to do so. Thus the very next week he reported to Huet that he too had done a public dissection. For Graindorge the most extraordinary thing about this session had been the large number of "ladies eager to see a heart unraveled in the style of M. Steno." This dissection also taught Graindorge an important lesson, however. He had seen Steno dissect hearts on two occasions, but Graindorge had never done it himself before he did it in public. He found himself somewhat embarrassed by the quality of his performance: "I was hindered because although you might be satisfied [you understand something like this] after watching it—seeing the demonstration done, talking the jargon, and doing the work are altogether different things. I acquitted myself so-so." Nevertheless, the public session pleased Graindorge. In what was obviously a response to the *assemblée* Bochart had hosted with Madame Luzerne the previous week, Graindorge had also arranged for the academy to perform its own dissection of a fox shark. Rather than doing the cutting himself, however, he let Busnel "visit the entrails" of this specimen. Graindorge did his dissection first, and although his dem-

[38]Ibid. Graindorge attributed both the skill and the ignorance to the fact that the man was a Protestant.

onstration ran over the allotted time, he was very pleased when Busnel proved able to "lay out [the organs of the shark] like they should be."[39]

It is difficult to assess Graindorge's performance at the first two sessions under his direction. Undoubtedly he was pleased, but it is easy to imagine that others may have taken less satisfaction than he did from the *expérience* with the oil of tobacco and his "so-so" performance with the dissection of a heart. Nevertheless, whatever his setbacks, he had made a start with his scientific program. It was proving more difficult than he had expected, but like Huet just a year earlier, Graindorge found that his skills in the laboratory began to improve markedly once he began to exercise them regularly. Over the next few months his demonstrations and dissections became much more successful.

By the middle of May, Graindorge had his first success with the efforts to engage other academicians in new research based on *curiosités*. In the same letter that reported the academy's public session to Huet, Graindorge also told his patron that Vaucouleurs had built a mechanism for weighing air. Apparently this project had been inspired by the description of such a device published in the new *Journal des Savants*. Like Huet and Graindorge before him, Vaucouleurs was having difficulties with his first *expérience:* "I do not understand anything from [the account in] the *Journal des Sçavans,* and M. Vaucouleurs who wanted to do the expérience from it found that over the course of several days the mercury was sometimes high, sometimes low—contrary to his expectations." The report they had read had not prepared them for this result, but barring some problem with their apparatus, Graindorge believed this *expérience* might open up exciting new possibilities:

> At least it should offer a principle by which we can recognize changes in the weather. It is certain that the air has weight. It follows that the quicksilver must rise [as the air takes on more weight], and that as the air becomes lighter the quicksilver should fall. We need [only] know when it takes on weight. I would have thought it would be in weather that is cold and rainy.[40]

Thus just as Graindorge thought the oil of tobacco demonstration had turned up something interesting when it failed to work, he found

[39]Ibid., 556, 550 (14 May and 13 January 1666).
[40]Ibid., 13 January 1666.

that Vaucouleurs's problems with weighing the air also presented a new area for research.

Vaucouleurs's device proved to be an extremely productive *curiosité*. Not only did Vaucouleurs keep track of its daily fluctuations but Graindorge managed to find various sources that Vaucouleurs used to find out about other investigations with the device. By late July, Vaucouleurs and Graindorge had learned that it should be called a barometer, and that some of the most interesting trials with such devices had been done by Blaise Pascal.[41] Pascal had confirmed Graindorge's opinion that differences in the weight of air should be attributed to how much it was "charged with vapors," but the observations Vaucouleurs had made did not agree at all with what was found in Pascal.

Pascal's "principles" had led him to explain the barometer as follows: "ordinarily," clear weather would make the mercury drop and cold and "charged" weather would make it rise. In other words, Pascal's general theory was the same as what Graindorge had "thought" two months before. Graindorge then explained that Pascal allowed for only one exception to this general rule for barometers: "However, he says that sometimes when the weather is improving it will rise and [sometimes] when it is overcast it will fall—for us, *this sometimes is the rule*. He saw that when [it fell during overcast weather], the clouds were thin and ready to dissipate soon. . . . This is contrary to our experience."[42] Graindorge and Vaucouleurs were puzzled by their results, but they continued to record their observations until Huet returned.[43]

Vaucouleurs's work with his barometer offers a perfect example of the kind of research Graindorge had called for in his letters from Paris. Vaucouleurs had done all Graindorge could expect. He had

[41]Ibid., 687 (26 July 1666). Graindorge got his information from Pascal's *Traité de l'équilibre des liqueurs*, which had been published posthumously in 1663. This work contained a short independent section titled "De la pesanteur de la masse de l'air." For discussion of these works and of Pascal's experiments weighing the air, see Michelle Sadoun-Goupil, "L'Oeuvre de Pascal et la physique moderne," *RHS* 16 (1963): 23–52; B. Rochot, "Comment Gassendi interprétait l'expérience du Puy de Dôme," *RHS* 16 (1963): 53–76. For a more general history of seventeenth-century experiments with the barometer, see W. E. Knowles Middleton, *The History of the Barometer* (Baltimore, 1964), pp. 38–82. Middleton discusses Pascal's *Traité* on pp. 44 and 48.

[42]BL, A 687 (26 July 1666); emphasis added.

[43]See, for example, ibid., 707 (20 December 1666): "Tout ce que nous savons a present ce n'est que ce qui le journal nous apprend et quelques observations du barometre qui continue touiours dans la mesme maniere quil a commencé contra la vray semblance, sabaissant en temps de pluye chargé et couvert."

located a *curiosité*, conducted his trials, and found a result that contradicted the best available opinion about nature. Yet neither he nor Graindorge knew what to do next. Graindorge displayed none of the excitement he had expressed over the failure of the oil of tobacco to kill the pigeon. Once again the academy had struck a blow against received knowledge and accepted opinion, but in this case the initial excitement of discovery seems to have paled against the long-term frustration of trying to explain what actually happened with the barometer. Graindorge's scientific program was working, but at times even he did not find it very exciting. That was a problem he had not anticipated.

During Huet's absence in 1666, the projects done by Vaucouleurs and Graindorge himself constituted the bulk of the work of the academy. The others—Villons, Lasson, Hauton, and Busnel—figure in Graindorge's reports only occasionally or in conjunction with a special session. The first time Villons is mentioned in any of Graindorge's letters, for example, is in September 1666, after he appeared for a session attended by visitors from Paris—Jean-Baptiste Du Hamel and Jean Picard, two men illustrious enough to lure him to the academy.[44] Lasson also appeared for this session, apparently the first he had attended since May. Most Thursday evenings, it seems, Graindorge and Vaucouleurs simply went to Huet's house and amused themselves as best they could with their own research. Nevertheless, they accomplished a good deal.

Following on his interest in Steno's dissections of the heart, Graindorge's first real research as an academician focused on the anatomy of the circulatory system. As early as May 1666 he began to concentrate particularly on the venous system, exploring not only the structure and location of veins but also their physiology. He wanted to know how they were constructed, what happened to them as the heart beat, and how they connected to the arteries; but above all he simply wanted to be able to trace exactly how blood flowed through them. In order to gain such knowledge, he injected various liquids in one vein—sometimes milk, sometimes water, sometimes wine—then opened another vein (or artery) and watched for this liquid to appear.[45] This *expérience* not only told him where the blood went, but also how fast it traveled and how much it "mixed" along the way.

[44]Ibid., 647 (27 September 1666). In December both would be among the fifteen men selected to become the academicians of the new Académie Royale des Sciences.

[45]Ibid., 689 (31 May 1666). Graindorge had learned these techniques at the Théve-not (VDLI, pp. 275–281 [3 June and 10 June 1665]).

Graindorge's letters offer little detail on what he actually learned, but it is clear that the work was excellent preparation for the blood transfusions he would begin doing in 1667. By the time transfusions became a *curiosité* pursued in London and Paris, Graindorge's knowledge of the venous system told him exactly how they should be done in order to achieve the desired results.[46]

Besides his interest in Vaucouleurs's barometer and his own work with the anatomy of the circulatory system, Graindorge also undertook new work on other *curiosités* during the summer of 1966. The most important involved the refraction of light as it passed between two media of different densities. Graindorge considered this topic important to his anatomy program. Any satisfactory explanation of vision required an understanding of how light passed from the air and through the eye. Thus refraction became the major topic of Graindorge's research throughout the remainder of 1666. His reports on his *expériences* with light are too long and detailed to allow a summary of their content here,[47] but it is important to note that he approached the question of refraction as a problem in *physique* rather than as a purely mathematical or theoretical question. He combed the literature he had available in search of theoretical statements and then set out to establish physical trials or *expériences* that would test their validity.

As always, Graindorge worked from a particular claim made by someone, in this case Isaac Vossius, and tried to find counterexamples that demonstrated errors. And as frequently happened, once he set to work he stumbled on interesting new *curiosités*. In this case, for ex-

[46]BL, A 1866, *inserti* 680, 679 (14 and 18 May 1667). Graindorge wrote these letters to Huet from Paris. In the first (14 May), he responded to Huet's inquiry about how to succeed with a transfusion by telling him the secret was first to bleed the specimen intended to receive the blood. Apparently the "trick" worked because in the second letter (18 May) Graindorge began by exclaiming: "A ce que je vois vous estes grand Maistre en matieres du transfusion." In other words, Huet had succeeded. The novelty of these transfusions seems to have worn off fairly quickly, however. Later (1669) Graindorge would report doing a transfusion for the royal intendant almost as if it were a parlor trick: "M. l'Intendant a envie de voir la transfusion. Jay fait chercher des suiets pour luy en donner le plaisir" (ibid., 594 [28 November 1669]). For further discussion, see Pierre Dionis, *L'Anatomie de l'home suivant la circulation du sang* (Paris, 1968).

[47]See Tolmer, *Huet,* pp. 340–348, for discussion of Graindorge's work, the treatise he wrote presenting his results (now lost), and the relation of Graindorge's *expériences* to Huet's own interest in vision. Questions relating to the anatomy of the eye and the optics of vision are among the topics most frequently discussed in the Graindorge correspondence.

ample, while trying to measure the refraction of light as it passed through a glass tube filled with water, he noticed that the surface of the water at the top of the tube was never flat, but that it crept up the walls of the tube, leaving a concave surface. He was at a loss to explain this phenomenon, and when he showed it to Du Hamel and Picard, neither could help him.[48]

During the summer of 1666 Graindorge was also seeking the causes of dew. This *curiosité* became important to Graindorge simply because Huet was interested in it. Graindorge had reported some speculations on dew while at the Thévenot the year before,[49] and Huet apparently found the subject intriguing. Sometime in late June 1666 Huet made an observation that he thought explained how dew formed. In the early evening he was sitting on the grass in a garden looking through his telescope. Just before sunset he set the telescope on the ground for a time, and when he picked it up again he found the underside damp and the lower half of the lens fogged. After sunset he found that both the upper and lower parts of the telescope became damp as it lay on the ground. These two observations convinced Huet that there were continuous exhalations of vapors from the earth, but that during the day the sun's heat dried them before any dew could form. He reasoned that as the sun began to set, it slowly lost its power to dry up the earth's exhalations, allowing the dew to form first on the lower sides of objects such as his telescope. He then set about trying to confirm his hypothesis.

Huet was not nearly so creative as Graindorge in designing *expériences.* The only other observation he found that supported his hypothesis was that dew formed first on the underside of leaves, but that after sunset moisture covered both the top and the bottom of a leaf. When he arrived at this point, he wrote a short treatise describing his observations and added some speculations on why heavy dew made for an unhealthy climate.[50] He buttressed these opinions with some alchemical lore he had learned from Hauton.[51] Huet then sent his

[48]BL, A 1866, 687, 647 (26 July and 27 September 1666). For a discussion of seventeenth-century theories explaining such phenomena, see E. C. Millington, "Theories of Cohesion in the Seventeenth Century," *AS* 5 (1945): 253–269.

[49]VDLI, pp. 272–274 (30 May 1665).

[50]As the vapors left the earth, they carried noxious particles with them.

[51]Huet's reference to Hauton's experiments with dew at the vernal equinox gives a clear indication that he was attending Huet's sessions immediately before Graindorge's return to Caen. For discussion of attempts to find special qualities associated with dew, see Theodore K. Hoppen, "The Nature of the Early Royal Society," *British Journal for the History of Science* 9 (1976): 1–24, 243–273.

treatise to Graindorge and asked for comments, reassuring him that he was not totally neglecting natural philosophy during his absence from the academy.[52]

Graindorge eagerly commented on his patron's latest work. Apparently this was the first sign he had had since April that Huet was still interested in scientific research. He gave Huet's observations some thought and then wrote a long reply, which is interesting for three reasons: (1) his obvious pleasure in the fact Huet had been doing research; (2) the way he translated Huet's simple "observations" into his own more formal *expériences;* and (3) the way he so easily assumed the role of Huet's scientific tutor. This letter provides a rare opportunity to read one of Graindorge's commentaries on Huet's work with a full knowledge of what Huet had actually sent him. His comments thus warrant quoting in some detail:

> I was awaiting your news with impatience. I am relieved that you can find a diversion so pleasant. Undoubtedly your spirit is better, and it is good you are giving it some freedom from that bond [the Origen project] that has occupied you for so long.
>
> The *expérience* where your telescope became damp on the underside during the day (with nothing appearing on the top) clearly shows that in a case where the heat of the sun dries the vapors, they cannot fall back [on top of things]. The second *expérience* [dew forming on both the top and the bottom of the telescope after sunset] shows the continuation of evaporation on the part of the earth and the commencement of the fall of vapors that have been raised. The third *expérience* [the way dew forms on the two sides of leaves] shows that equal amounts of vapor rise and fall. If you had continued [with this research] perhaps you might have found how much [vapor] the earth loses and how much the air gains. This might be shown with two large bell jars of glass with alembics; the one would have its opening turned to the ground and the other toward the sky.[53]

Graindorge thought these two jars might collect different quantities of dew, which would demonstrate how much of the evaporation from the earth was resettling during the night. Rather than comment on Huet's medical and alchemical speculations, he turned the discussion to the usefulness a knowledge of dew might have for agriculture. It appears unlikely, however, that either he or Huet ever attempted any more trials to capture dew. After sending his treatise to Graindorge,

[52]The text of Huet's treatise is presented in its entirety in Tolmer, *Huet,* pp. 345–346.

[53]BL, A 1866, 688 (13 July 1666).

Huet undoubtedly considered the subject closed. For Graindorge, no topic was ever completely exhausted; he simply did not have time to follow up every *curiosité* he found.

Graindorge possessed seemingly limitless curiosity about the marvels of nature. Even as he was pursuing these major scientific interests as part of his work for the academy, he also kept himself amused with minor projects. While Graindorge was at his estate in Villons during the harvest, for example, Huet wrote to tell him that moles have eyes. Graindorge doubted this claim and immediately set out to discover if it was true, and what effects (if any) a mole's underground life had on the basic anatomy of its eye. He also became interested in insects during the harvest. He sent back to town for books, and to satisfy his curiosity more directly, he captured a wasp and a black caterpillar. He kept them both in jars for a time, feeding them and trying to decide what their "business" was. The wasp he then dissected; the caterpillar he kept in order to discover what kind of butterfly would emerge from its cocoon. He was amazed by what he discovered when he dissected the wasp, and equally surprised when one of his servants told him the caterpillar would pass the entire winter wrapped in its cocoon—without food.[54]

Overall, Graindorge's letters to Huet offer ample proof that he was personally able to put his new ideas into practice when he returned to Caen. It seems that wherever he turned, he found *curiosités* that led him to new research. Whenever he began an *expérience,* he found new knowledge, frequently knowledge that was inconsistent with accepted views about nature. Graindorge's letters, however, also offer overwhelming evidence that the Académie de Physique was not "animated" by the success of his program as it had been under Huet. Apart from his reports on Vaucouleurs's trials with the barometer, Graindorge's letters describe not a single sustained piece of research done by any other academician after Huet's departure. Graindorge had planned to make his *curiosités* the focus of the academy's sessions. He had found his *curiosités,* but they had not brought the academicians to Huet's house. In short, although Graindorge succeeded with the science, he failed with the academy. No matter how successful his and Vaucouleurs's *expériences,* the term "académie" lost its meaning if they were the only ones who came to sessions.

[54]Ibid., 560, 668, 664 (6 September, [October], and 29 November 1666). Graindorge described the wasp as "une bête armée de pied en cap" and was impressed by its seeming ferocity; he concluded his long account by claiming, "Jamais je n'ai rien vu d'aussi félon" (VDLI, pp. 324–327 [October 1666]).

THE LESSON OF 1666

The lesson of 1666 was obvious to Graindorge, and his perceptions warrant attention: long-term success with the Académie de Physique required the active involvement of its patron/*chef*. Even as he had faced the prospect of "renewing" the academy in the spring of 1666, Graindorge knew it would be difficult without Huet. In early May his "successes" with the oil of tobacco and the dissection of a heart gave him some confidence and new hope. Such optimism was short-lived, however. Before May was out, the academicians had stopped attending sessions. When Graindorge found that not a single one was coming to the session he had scheduled just two weeks after the public dissections, he told Huet: "I greatly fear that your absence has broken the neck of all our wonderful plans."[55] His fear was justified.

Through the summer his letters show nothing to indicate the situation had improved. Graindorge was enthusiastic as he reported his own research, but his letters offer no suggestions of group activity. In September the presence of Du Hamel and Picard gave the academy new life, but with their departure the academicians once again stopped attending. From then until Huet's return, Graindorge seems to have given up his efforts to "animate" the academy. In fact, for the first time in his correspondence with Huet, Graindorge's letters began to deal with a variety of subjects other than science.[56] As for news of the academy, he described his own work and Vaucouleurs's continuing trials with the barometer, but said nothing to indicate that any of the others ever came to a session. By December, Graindorge was uncharacteristically pessimistic and depressed: "I dare not think it, but I do—our académie physique is destroyed." Only the news that

[55]BL, A 1866, 648, 541, 548, 557 (10 March, 10 and 19 April, and 21 May 1666).

[56]During this period, for example, Graindorge's responsibilities as an *échevin de ville* required that he give a great deal of time to the problems caused by the fourteen companies of infantry quartered on Caen for the winter (ibid., *inserti* 666, 665, 664, 707, 663, 561 [1 November 1666–10 January 1667]).The continuity of his discussions of the problems with the "soldats" and "gens de guerre" helps to date the series of letters. Graindorge was personally responsible for the care of four companies (1 November 1666). Although he did not realize it at the time, this extraordinary military buildup during the winter of 1666–1667 was part of the preparations for the War of Devolution, which Louis XIV would launch by invading Flanders on 24 May 1667. The chronology of this war takes on considerable significance in the dating of a number of Graindorge's letters during 1666 and 1667. In turn, a precise knowledge of the sequence of Graindorge's letters is essential for understanding both the scientific and organizational development of the academy. For chronological details of the planning, military buildup, and course of the war, see Ernest Lavisse, *Louis XIV* (Paris, 1978; rpt. 1900–1912), 2:104–113.

Huet was planning to return sometime in January renewed his hope.[57]

The news of Huet's imminent return aroused Graindorge's enthusiasm for the academy once again. Combined with the onset of extraordinarily cold weather, this enthusiasm led him to open up an entirely new line of scientific research. The terrible weather gave him the opportunity to begin a series of *expériences* on the effects of cold. He tried to obtain a copy of Robert Boyle's new work on the subject, and by early January 1667 he was already planning new *expériences* based on what little he knew about Boyle's work. Obviously looking forward to Huet's return, Graindorge issued a challenge to his patron: "The other day I read that someone had fashioned two glass balls in such a way that when they were put in cold water one went to the bottom and the other to the top. Adding hot water, the ball at the top sank, and the one at the bottom rose to the top. . . . Redden your fingernails a little seeing if you can lay bare this mystery."[58] Evidently the prospect of having Huet back in Caen had put Graindorge in a jocular mood.

In January 1667 Graindorge could afford to indulge in some lighthearted humor. Not only did he know that his patron was returning to Caen, but he also knew that Huet had already resolved to establish the Académie de Physique on a firm basis as a research society. Huet had even told his own patron, Jean Chapelain, about this plan.[59] Huet's presence in Caen during 1667 would have precisely the effects Graindorge hoped for. His active involvement did animate the academy. As Huet and Graindorge were now together, their correspondence was suspended for much of the year,[60] but there is ample

[57]BL, A 1866, 707, 663 (20 December and last week of December 1666).

[58]Ibid., 561 (10 January 1667). Evidently Graindorge had become frustrated in his efforts to obtain Boyle's *New Experiments and Observations Touching Cold* (London, 1665) and was also having difficulty finding someone capable of translating the work. He blamed the problem of securing a copy on the continuation of the maritime war with England.

[59]Chapelain to Huet, 25 January 1667, BL, A 1866, 264. In what was obviously a response to Huet's announcement of his intentions concerning the academy, Chapelain wrote: "Je suis bien aise au reste que vous allies recommences vos exercices philosophiques et que vostre bon courage vous fare resoudre a prendre sur vous la despense necessaire à la descouverte des Mysteres de la Nature. Que s'il vous arrive de trouver quelque chose de solide et de considerable en ce genie la, j'auray beaucoup de joye d'en avoir communication pour la donner au Journal des Sçavans ou a l'Académie Royale."

[60]Graindorge made a trip to Paris and Compiègne between the end of April and the end of July 1667. Nine letters from that period have survived, but this series appears to be incomplete. Huet would leave Caen again in October, at the time arrangements for the academy to be taken under royal protection were being finalized.

evidence of the effects of Huet's presence on the academy. In May, Huet succeeded with a public demonstration of the transfusion of blood. Graindorge, who happened to be in Paris at the time, was so pleased that he proclaimed Huet the "Grand Master of transfusions." During the summer Huet and Graindorge had enough confidence in the prospects for the academy to try to lure Jean Picard away from the new Académie Royale des Sciences with the promise of a post at the university in Caen.[61] In September 1667 the first reports of dissections in Caen reached Henry Oldenburg at the Royal Society in London.[62] By October the academy had expanded from six to ten academicians,[63] and they had formalized the sections for *mathématique* and *physique* with five members each.

Huet left Caen during October, but before he went, negotiations that were to place the academy under the protection of Guy Chamillart, the royal intendant of Lower Normandy, were already under way. Within weeks the Académie de Physique would have the *approbation* of Louis XIV and the promise of funds from Colbert. Huet had revitalized the group, and Graindorge was thrilled. By late 1667 it seemed the academy he had envisioned for two years was about to become a reality.

Given Graindorge's belief in December 1666 that the Académie de Physique was "destroyed," Huet's ability to reanimate the group dur-

[61]BL, A 1866, 679, 680, *inserti* 562, 563, 676 (14 and 18 May; 4, 14, and 19 July 1667). In July Graindorge recounted his attempts to open negotiations (in Huet's name) to bring Jean Picard to Caen. After failing with Picard, who was reluctant to leave Paris and the Académie Royale for the dull routine of a provincial teaching post, Graindorge discussed various other candidates he thought might please or displease Huet. What is extraordinary about Graindorge's reports is the degree of influence he assumed he and Huet would be able to exercise in this affair. For example, on 19 July he reported not only his ideas on the kind of candidate they should seek but also his conversations with the royal intendant of Lower Normandy and the bishop of Bayeux on the subject. He thought the vacant chair at the university should go to someone who was proficient in hydrography, mechanics or fortifications, and chemistry, and told Huet: "Jen ay parlé un peu a Mr de Bayeux et un peu a Mr l'Intendant qui est icy qui na pas conclu non plus qua son ordinaire. Jay rencontré de hazard Mr Picard ce matin qui ma dit quil cherche un homme qui nous fust propre. . . . Mr l'Intendant ma parlé dun capucin. Jugez quel home et sil est bien propre a un tel dessein dordinaire ignorante et presomptueux."

[62]In a letter dated 17 September 1667, Oldenburg wrote to Robert Boyle summarizing the news contained in the first letter from abroad he had received since his release from prison. Among the items reported is that "at Caen they have dissected the eye of an owl; wch also I expect the particulars off" (*CHO*, 663).

[63]BL, A 1866, 675 (31 October 1667). In this letter Graindorge reported having accepted M. Savary to fill "la dixieme place."

ing 1667 proved remarkable indeed. In less than a year Huet not only had rebuilt the original group but had enlarged it and was preparing to entrust its future to the patronage of the monarchy. Certainly his influence was dramatic; so much so, in fact, that it becomes impossible to deny the validity of Graindorge's assertions about Huet's importance to the academy. The source of Huet's influence lay in his role as patron/*chef* rather than in his skill as a scientist.

The entire concept of the Académie de Physique as a research institute belonged to Graindorge. It was an idea he had persuaded Huet to support. Moreover, Graindorge was clearly the one who was adept at finding *curiosités* and designing new *expériences.* The academy's success with blood transfusions during 1667, for example, was based on Graindorge's expertise, not Huet's.[64] Moreover, Graindorge's commentary on Huet's treatise explaining dew gives ample reason to believe that Huet not only accepted Graindorge's superior scientific expertise but took it for granted. The scientific program belonged to Graindorge; Huet's role was that of patron. That arrangement, however, presented the two of them with a problem.

Huet's absence during 1666 had shown Graindorge that he had to have his patron by his side in order to animate the academy. That lesson could only have been reinforced by the successes of 1667. Huet was willing to help Graindorge. Yet the very fact that by the end of the year Huet was preparing to turn his academy over to the royal intendant indicates that he was unwilling to extend his commitment beyond the point at which the group had purpose and momentum sufficient to carry on without him. By late 1667 he had the Origen manuscript in press and was ready to put his own career on a larger stage than the one available in provincial Caen. Graindorge, for his part, had indeed learned a valuable lesson during 1666 and 1667. By the time Huet left, he knew that creating the *siècle-d'or* would take more than a frontal assault on the mysteries of nature. He knew that before any such attack could begin, he first had to bring the academicians into their laboratory.

As Graindorge faced Huet's departure in October 1667, however, he had every reason to believe that his problems were all but resolved. His letters show none of the anxiety he had expressed in the spring of 1666. He had changed none of his ideas about what the academy should become. On the contrary, he was more committed than ever to his ideas about research and the importance of coordinated efforts.

[64]See n. 46 above.

The difference was that, with the prospect of Chamillart taking over as the academy's patron, he thought he had the means at hand to keep the academicians involved in their research. Even though academicians had stopped attending as soon as Huet left, Graindorge was satisfied to carry on with the scheduled sessions until the intendant was ready to take over the academy. Indeed, he was determined to do so, even if he had to work alone:

> Thursday I was at [my estate] in Villons and I had for our academy a kind of screech-owl that was badly injured. . . . I ate early so I could leave.
>
> Well, the weather was horrible, with such a heavy rain . . . that it was impossible [to keep dry] on the road. I suffered every misery possible. I took my bird, [however,] and dissected it at my leisure. Here is what I found. . . .[65]

Though no one else had bothered to come to the session, Graindorge was far from distressed. At the time, he did not realize that the session would be the last officially under Huet's patronage, but he knew the academy was about to be taken under Chamillart's protection. He thought he was finally going to get the patron/*chef* he needed.

Thus, as November 1667 approached, Graindorge was enthusiastic about the academy's future. Before that month was out, he would be much less so. The intendant Chamillart wanted a great many changes in the Académie de Physique. Not all of them would please Graindorge. In order to understand why, however, we must first deal with the conditions laid down when Huet turned the academy over to Guy Chamillart, the king's representative in Caen.

[65]BL, A 1866, 675 (31 October 1667). Graindorge then provided Huet with a detailed description of the dissection he had done on the owl's eye. Huet considered the description important and passed it on to Henri Justel. Justel in turn described Graindorge's dissection of the eye in a letter to Henry Odenburg in December 1667, and Oldenburg passed it on to Robert Boyle on 17 December (*CHO*, pp. 721, 728). Thus, despite the notoriously poor communications of the day, news of Graindorge's dissection had passed from Caen to Rouen, from there to Paris, from Paris to London, and from London to Oxford in just over five weeks. This was just the kind of dissemination of scientific knowledge Graindorge considered vital.

4

The Royal Incorporation

By the fall of 1667 Graindorge and Huet had had five years' experience organizing scientific activity. Their academy had enjoyed just two genuinely productive periods during that time. The first came while Graindorge was in Paris, the second during 1667. Those experiences had taught them a simple formula for success. Both productive periods derived from the empirical scientific program Graindorge had learned at the Thévenot; thus the importance of that program was obvious. More important, however, Graindorge's program had become productive only when Huet was personally acting as the academy's director, or *chef.* More than anything else, then, Huet's willingness to fill that role had made the academy function. To Graindorge and Huet, the academy's most pressing need was also obvious. Even judged by the minimal criterion of survival, success depended on the patron's physical presence in the laboratory. Otherwise there was no academy.

Identifying what worked was a simple task; ensuring its implementation proved more difficult. By the fall of 1667 Huet was finally ready to publish the *Origenis Commentaria,* and that process was going to require extensive absences from Caen.[1] Moreover, as the *Commentaria* attracted attention within the Republic of Letters, Huet might well decide it was time to shake off the dust of provincial Caen and move his career onto the larger stage of Paris. And indeed he did. In the fall

[1]Brennan, "Culture and Dependencies," discusses how and why the publication of that work kept Huet in Rouen and Paris (pp. 111–115).

of 1667 Graindorge could not possibly foresee that over the next five years (sixty months) his patron was going to spend less than ten months in his native Caen.[2] Nevertheless, even without such precise knowledge, Graindorge realized that he could not count on Huet's continued presence. He knew there was a problem. The academy needed a *chef* who attended sessions. Huet agreed, and before he left Caen in October 1667 he had entered negotiations to place the Académie de Physique under a new director, Guy Chamillart, the royal intendant in Lower Normandy.[3] Graindorge was thrilled. Surely this man could galvanize activity.

Chamillart certainly did have an impact on the group. When Huet left Caen in October, the Académie de Physique was a private-patronage scientific organization. Just three months later it had become a royal academy. A great deal happened in that time. In November and December 1667 the academy reorganized under its new *chef,* Chamillart. Then, in January 1668, Chamillart obtained three extraordinary marks of distinction for his group. First he secured Louis XIV's personal recognition for the academy as a royal institution, then the pledge of scientific coordination from the Académie Royale des Sciences, and finally the promise of royal funding from Colbert. In less than ninety days Chamillart had taken the group from the micro level of Huet's private patronage to the macro level of state-supported institution. He made the Académie de Physique a royal corporation.[4] Nominally it had become one of the preeminent scientific institutions in Europe.

The events of late 1667 and early 1668 made the Académie de Physique the French monarchy's second royal scientific academy. For contemporaries, its elevation to this status just a year after Colbert created the Académie Royale des Sciences suggested that the French monarchy had committed itself to a major program of state-

[2]During the sixty months from November 1667 through October 1672 (the period for which we can meaningfully discuss a royal academy of sciences in Caen), Pierre-Daniel Huet was absent from Caen for fifty-one. The only times he was in his native city were between mid-March and mid-August 1668 and from mid-May through September 1670 (BN, Fr, n.a., 1197).

[3]For discussion of the French intendancies, see Charles Godard, *Les Pouvoirs des intendants sous Louis XIV, particulièrement dans les pays d'élections, de 1661 à 1715* (Geneva, 1974 [1901]).

[4]Although not focused on France or French history, Lon Fuller's *Legal Fictions* (Stanford, Calif., 1967) offers the surest introduction to the legal concept of incorporation. The Académie de Physique's incorporation was simply a special case of the more general theory Fuller discusses.

supported science. In England, for instance, Henry Oldenburg told Robert Boyle that the Académie de Physique's reorganization signaled Louis XIV's desire to create Europe's first large-scale scientific research capability. To all outward appearances, Chamillart's reorganization gave the Académie de Physique every promise of a grand and productive future.

Behind those appearances, however, the reality of royal incorporation fell far short of expectations. Most important, all the marvels Chamillart, Colbert, and the king produced did absolutely nothing to resolve the basic issue Graindorge and Huet had already identified as the academy's first concern: who was to act as *chef*? Not the king or Colbert, certainly. And when Chamillart himself proved unwilling to fill the role, Graindorge found the academy in no better position than it had enjoyed under Huet. Indeed, by the end of January 1668 the academicians had proven themselves just as unwilling to work for an absentee Louis XIV, Colbert, or Chamillart as they had been to work for an absentee Huet.

To understand the Académie de Physique's royal incorporation requires us to deal with one basic fact: incorporation originated in an effort to ensure the group's organizational stability but did virtually nothing to accomplish that end. If anything, the academy found itself in a more precarious state following royal incorporation simply because all the royal "favors" gained created greater expectations for success. After January 1668 the academy's continued existence was no longer a simple matter of personal and local concern. Once Louis XIV recognized this academy as his own, any failure would tarnish his *gloire*. Even if such failure did only minuscule damage to the Sun King's reputation, such men as Graindorge, Huet, and Chamillart (and even Colbert) wanted to avoid responsibility for any embarrassment to Louis XIV. After all, this was a king with whom courtiers banked the difference between a royally bestowed smile and a frown. Royal incorporation thus produced a paradox: it did almost nothing positive to ensure success, and it dramatically raised the stakes for failure.

Royal incorporation produced a crisis for the Académie de Physique. In essence, it was a crisis of patronage: how could the men who created this royal institution ensure its organizational future? Of course, that question casts an ironic tone on the academy's entire history. Ensuring organizational stability was the problem Huet and Graindorge had tried to solve when they accepted Chamillart as the group's new *chef.* But Royal incorporation turned the problem of patronage into a crisis of patronage. How could such a thing happen?

Ultimately the explanation lies in the complexities of the patronage system governing the Académie de Physique. We have already seen some aspects of this system in the personal relations between Huet and Graindorge as well as in Graindorge's inability to function as the group's *chef* during 1666. Nevertheless, in order to understand why the academy's royal incorporation was its undoing, we need to analyze the workings of the academy's patronage even more closely. Everyone concerned agreed that the key to the academy's future lay in establishing the proper system of governance under something we can only call "patronage." Unfortunately, almost no one agreed on what that mechanism should be. Curiously, this lack of consensus did not spring from any basic differences over what patronage implied. Everyone knew the academy had to be governed in a complex, multitiered system involving intellectual, financial, legal, and operational functions. Moreover, all agreed that these roles had to be parceled out among those directly responsible for the academy: Chamillart and Graindorge as well as Colbert and Huet. Disagreement came only when it was time to establish exactly what the first priorities among these responsibilities were, or when it was time to decide exactly who should exercise each. At this point we can turn to a closer consideration of how such issues worked out in practice.

CREATING A ROYAL ACADEMY OF SCIENCES

Graindorge knew by late October 1667 that the Académie de Physique was about to undergo its second major reorganization. Despite Huet's absence and the fact that Graindorge found himself alone at the session scheduled for the last week of October, the letter in which he renewed his correspondence with his absentee patron was full of optimism and enthusiasm.[5] Chamillart was about to become the organization's *chef.* Unfortunately, neither that first letter nor any of his subsequent correspondence gives any direct evidence on who initiated the discussions leading to this change, but Graindorge's letters do clearly indicate that the project was under way and that the issue was in negotiation. The genesis of the project is less significant, however, than the fact that by the end of October all concerned were

[5]BL, A 1866, 675 (31 October 1667).

enthusiastic about the prospects of Chamillart's association with the group. All parties—including Huet, Graindorge, the academicians, and Chamillart himself—had reasons to look forward to this new arrangement. Everyone was eager for the last details to be worked out so that the academy could begin to shape its new work.

Unfortunately for everyone concerned, Huet's absence made Chamillart's takeover somewhat more time-consuming and cumbersome than necessary. Chamillart and Huet both wanted to formalize the transition in a document that would become the academy's statutes. As October ended they were trying to construct such a document through the mail. Apparently the arrangement proved unsatisfactory. At the point where Graindorge's letters pick up these negotiations with detailed information (the first week of November), he was just beginning to act on Huet's instructions in trying to finalize the agreement with Chamillart. By then a draft of the statutes did exist. It was in Chamillart's possession, and presumably the intendant had written it. Huet had seen the document, or at least a copy, because Graindorge possessed a list of new demands that Huet wanted made in these last negotiations. On the morning of 3 November 1667 Chamillart invited Graindorge into his office and told him that Huet was ready to allow the two of them to finalize the terms of the statutes. Armed with Huet's list, Graindorge sat down to the task.[6] The results delighted Graindorge. Chamillart acceded to every one of Huet's demands and even allowed Graindorge himself to write several new articles into the statutes. By the time they finished, Graindorge and Chamillart had produced a document that spelled out exactly what the "new" Académie de Physique was to become. They had defined an organization composed of two sections (one for *la mathématique* and one for *le physique*) and had specified that the group was to work with astronomy, chemistry, and natural philosophy. Each section was to have five academicians, and the statutes listed the men (including Huet) who would fill the ten chairs. The document named Graindorge as secretary and, separate from the ten academicians, Chamillart as the organization's *chef.* Furthermore, the statutes mandated that the academy meet at Huet's house: Huet had promised a room for general meetings, a laboratory, and space for an herb garden. In addition, the statutes named the surgeon Busnel as the academy's anatomical demonstrator while also providing for a new laboratory technician. Finally several articles described exactly how the academy

[6]Ibid., 564 (11 November 1667).

should function, specifying such particulars as restrictions on admission, the qualifications for membership, and statements on working procedures.[7]

Subsequent events at the academy make it absolutely certain that as Chamillart and Graindorge sat down to finalize these articles for the statutes, both understood that they were drawing up a legal document, a form of contract. Unfortunately, they failed to reach any sort of understanding on who was bound by which terms in their agreement. Graindorge certainly thought the intendant had pledged himself to fulfill these statutes; this misapprehension explains, for instance, why he was so pleased when Chamillart allowed him to write the requirement for subscriptions to the *Philosophical Transactions* and the *Journal des Savants* into the document. Chamillart just as surely felt that Graindorge pledged the academy's membership to undertake all the marvelous endeavors spelled out in the document. Little wonder he was so free in granting demand after demand as Graindorge put them forward.

According to the report Graindorge sent Huet, he had only two disagreements with Chamillart in the course of marking up the academy's statutes. One concerned something that may have seemed to loom large in early November 1667 but that proved to be a minor matter: a restriction on the admission of visitors to working sessions. Both Huet and Graindorge wanted the academy to work behind closed doors. They wanted to admit only those visitors who possessed scientific credentials; Chamillart wanted more open access. In the end, they compromised with a careful wording of the article in question. They would admit "nonacademicians otherwise eminent in quality and merit."[8] At first, such wording may make it seem that Graindorge had won the argument for scientific professionalism. That was not the case, however. In seventeenth-century France, "otherwise eminent in quality and merit" would certainly admit any scientist to this academy, but it was also an open invitation to any *grand personnage* who might show an interest in the marvels of science. Worded as it was, the provision was virtually meaningless.

The second issue Chamillart and Graindorge debated may seem equally mundane at first, but it concerned something truly fundamental to the difference between the way these two men saw what they were doing. They agreed that the academy needed a new name,

[7]Ibid.
[8]Ibid.

but not on what the heading for the academy's statutes should be. Graindorge wanted something straightforward and descriptive, and thus proposed to title the statutes *Institution de l'Académie Physique par M. de Chamillart.* Chamillart had his own ideas about a straightforward, descriptive title: *Institution de l'Académie Royale par le Roy Très Chrétien Louis XIV.* Graindorge objected on the grounds that use of the king's name for the statutes required the issuance and registration of letters patent.[9]

Graindorge's letters do not reveal how they actually settled the dispute about the academy's official name, but again and again over the course of the next several years his letters demonstrate that he was sadly mistaken both in proposing his choice of a name and in objecting to Chamillart's. The intendant had no need to follow some ritualistic legal formula in creating a royal academy. Even as these men worked on amending the statutes that Thursday morning, they drafted the articles for the academy's royal incorporation.

It is almost inconceivable that Chamillart ever entertained the notion of obtaining royal letters patent for the Académie de Physique. There was no need for any such legal instrument, but even more important, to have presented such a document for registration in a *parlement*—whether of Paris or of Rouen—would have been politically foolish. This is a point that needs underscoring before we can fully understand the intendant's dealings with the academy. Chamillart had at least two good reasons for avoiding an attempt to register letters patent that the academy simply did not need.

For many contemporaries, Chamillart's involvement with this provincial academy was problematic: he was the royal official charged with sinking the monarchy's bureaucratic roots deeper into Lower Normandy, a notoriously fractious province where memories of brutal repression of revolts against royal authority still lived.[10] Any number of parties might have eagerly used a legal battle over an academy's letters patent as a forum in an attempt to limit the intendant's powers. For the Caennais of the 1660s, the presence of a royal intendant in their midst was not yet something taken for granted. The "modern" institution of the intendance dated only to 1636. Moreover, since the first appearance of the intendants, relations with these royal officials

[9]Ibid.

[10]For an analysis of the situation in Normandy following the revolt of the *Nu-pieds*, see Roland Mousnier, *Peasant Uprisings in Seventeenth-Century France, Russia, and China*, trans. Brian Pearce (New York, 1970), pp. 87–113.

had been extremely troubled.[11] During the monarchy's financial difficulties of the late 1630s and 1640s, the intendancy served as an instrument for extraordinary fiscal extractions and the repression of revolt. Clearly that was the intendancy's original purpose. Then, during the first Fronde, the *parlementaires* had forced the monarchy to abolish the office in Normandy—a province where antiroyal sentiment ran high. During the Princes' Fronde, the governor of the province went into open rebellion. Following the Fronde, no intendants appeared in Normandy again until the late 1650s. And then problems still plagued the office. Chamillart, who took up his post only in 1666, was in fact the first effective, full-time intendant in lower Normandy following the reestablishment of the office. In effect, after a thirty-year history of difficulty with the intendance, Chamillart was the first royal intendant ever charged with "normal," day-to-day administration of Lower Normandy. Any number of parties could (and did) have reason to oppose him if they were given an opportunity.

If those general objections to the intendance were not enough to create opposition to Chamillart, he exhibited another salient characteristic that made him an obvious target for *parlementaires* interested in challenging royal power. Guy Chamillart was a political *fidèle* to Colbert, the minister who was just then emerging as a symbol of the new royal ascendance. Colbert's own rise to prominence had come through the destruction of the powerful Fouquet, whose cause continued to enjoy support among *parlementaires*.[12] It is almost inconceivable that a man in Chamillart's political position would have contemplated trying to formalize the Académie de Physique's existence in any *parlement*. Colbert himself was careful in his dealings with these bastions of corporate privilege in the 1660s and sought registration for letters patent for a new academy only once during the decade.[13] In most cases, there was simply no need for such documents. For the Académie de Physique, there was surely no need. Both the Academy

[11]Yver, "Le Gouverneur et les premiers intendants," pp. 320–359; Mousnier, *Peasant Uprisings*, pp. 87–113; Godard, *Pouvoirs des Intendants;* Brennan, "Culture and Dependencies," pp. 14–34.

[12]See Lavisse, *Louis XIV*, 1:79–88, 182; Meyer, *Colbert*, pp. 159–181; Murat, *Colbert*, pp. 89–112; Wolf, *Louis XIV*, pp. 136–144; *Colbert, 1619–1683* (Paris, 1983 [exhibition catalogue]), pp. 37–44.

[13]Colbert obtained letters patent for his reorganization of the Académie des Beaux Arts, but among his new foundations during the 1660s only the Académie de France de Rome received letters patent. For discussion and documents, see Léon Aucoc, *Lois, statuts, et réglements concernant les anciennes academies et l'Institut de 1635 à 1889* (Paris, 1889).

of Inscriptions and the Académie Royale des Sciences, for example, already existed without letters patent.

Chamillart never spelled out to Graindorge his plan for the academy's incorporation—or, if he did, Graindorge failed to understand what it entailed. When he claimed victory over Chamillart in negotiating the statutes, Graindorge clearly thought the intendant was muddling about. That was a misperception. With the aid of historical hindsight, we can look at Graindorge's reports on Chamillart's behavior and see the intendant acting with a consistent clarity of purpose, even in marking up the statutes. Chamillart never intended to become the academy's *chef* (or its patron) in the sense that Graindorge wanted. Chamillart was a royal administrator and always behaved like one. Graindorge thought Chamillart was being grandiose and presumptuous as he insisted that the academy was an "institution . . . par le Roy . . . Louis XIV." In reality, Graindorge was being naive and obtuse when he failed to understand that that was exactly what the academy became the moment Chamillart became its *chef*.

In marking up the final copy of the academy's statutes, Chamillart created an administrative document for use in Paris. To use an anachronistic term, he served as a government contracting officer. For him, the Académie de Physique was a viable private-patronage organization, with a membership eager to serve the monarchy. In recognizing the organization's potential, he acted as the king's agent in making incorporation possible. The first step in that direction involved getting the members of the academy to specify exactly what special capabilities they promised to provide the king. All too clearly that is what Chamillart—the royal administrator—thought Graindorge was offering with his amendments to the statutes. To him, every new amendment spelled out the organization's structure and purpose more clearly. Each new clause made the contract easier to administer from Paris. Every article added to the statutes clarified the academy's corporate responsibilities.

Chamillart planned to incorporate the academy through ministerial action and administrative procedure. This was, in fact, the same approach to royal incorporation Colbert had already used with both the Academy of Inscriptions and the Académie Royale. The statutes completed his first step in that direction. With that document in hand, his next step was to charge the academicians with formulating specific research projects to fulfill the organizational form defined in the statutes. Since the meeting with Graindorge took place on a Thursday morning, his opportunity to take that next step came later the same day. Unfortunately, as Huet was in Rouen, the academicians had not planned to attend the session scheduled for that evening, and in the end Graindorge could notify only four other members of the academy that they had a new *chef.*[14] Chamillart thus met his new academy at half strength. The session proved an inauspicious beginning for his relationship with the group.

When Graindorge called at Chamillart's house to escort him to the academy's meeting, he found that the intendant had already invoked the double meaning of the article that allowed nonacademicians to attend sessions. He had invited two guests to the new royal academy. One was M. Montatère, a *grand personnage* who "will not fail to tell the king that he attended our session." The other was a scientist, a young doctor who claimed to have a method for sweetening seawater. As soon as the academicians approved the statutes, Chamillart wanted them to give an expert opinion on the feasibility of this desalinization project. The young doctor's handiwork did not impress Graindorge, but he recognized Chamillart's interest in it. Thus he led the academicians through an evaluation, concluding that the device was impractical: too slow, too cumbersome, too expensive, and the water it produced tasted foul.[15]

The academicians wanted to show off their own expertise to the intendant and his visitors. To that end, Graindorge led them in a discussion comparing the vision of owls and cats. Obviously he was trying to base the discussion on the dissection he had done just a week earlier. Since none of the other academicians had attended that session, however, no one had seen his dissection of the owl's eye. Apparently the discussion was less than enthusiastic, but Graindorge pressed on with it: "We amused ourselves as much as possible with this talk."[16]

Finally the time came for what Chamillart wanted to discuss: plans for new projects. The academicians suggested that they design and perfect new scientific instruments. They thought the construction of a thermometer and a hygrometer as well as further trials with the academy's barometer might prove worthwhile. Those were "dubious

[14]Those in attendance included Graindorge, La Ducquerie, Vaucouleurs, Cally, and Chasles. Absent: Huet, Lasson, Villons, Savary, and Hauton (BL, A 1866, 564 [11 November 1667]).

[15]Ibid.

[16]Ibid.

projects in the opinion of the intendant." He wanted them to begin draining the swamps in his generality and to build pumps that would allow construction of fountains in the city's public squares. Since those projects were expensive and time-consuming, however, he also suggested that the academicians could make themselves useful immediately by fashioning a weathervane for his house. He wanted it to look like the one on Colbert's house in Paris. Graindorge countered Chamillart's suggestions for such "useful" projects and proposed instead that the academy build terrestrial and celestial globes, which had the double advantage of "great beauty as well as utility."[17]

Besides the implicit differences between the kind of work Chamillart wanted and what Graindorge and the academicians were willing to produce, an even more fundamental problem underlay this debate over a new research program. For the second time that day Chamillart and Graindorge failed to bridge the gap between their visions of the intendant's role in the academy. Chamillart had not yet said a word about how he planned to finance the grandiose proposals he made. Indeed, even as he talked, it became painfully clear that he felt no personal obligation to finance the academy at all.

Graindorge and the others knew that Chamillart wanted them to expand the academy's program. They even recognized that the intendant wanted public-works projects. As *chef*, he had every right to ask for such things. Nevertheless, they expected a *chef* to serve as their private patron, someone who paid the costs of any special new projects he wanted done. Chamillart, however, acted as if he expected the academicians to pay for everything. Graindorge and the others reacted as if they had been betrayed. Vaucouleurs voiced the opinion of the assembled group when he told Chamillart that before the academy could undertake anything new it needed both a general operating fund and specific funding for individual projects.[18] As long as the intendant pushed for his public-works projects without funding them, Graindorge and the other academicians would continue the program of dissections and their work with scientific instruments.

When the meeting of 3 November 1667 adjourned, Chamillart had indeed taken his second step in carrying out the academy's royal incorporation. He had delivered the charge to formulate new projects. As far as he was concerned, he had discharged his responsibilities as the academy's royal administrator. In that role he had

[17]Ibid.
[18]Ibid.

nothing further to contribute until the academicians gave him a response. After this first session, he did not even check on them. For the next six weeks he simply left the group alone to reorganize its program.

For their part, the academicians immediately took a stand that they maintained throughout their dealings with Chamillart and the monarchy—no guaranteed funding, no royal projects. Until Chamillart demonstrated his ability to act as patron, they refused to work on a single project he urged. In that first face-to-face confrontation with Chamillart, the academicians argued that they could not afford the academy's projects. They held to this position steadfastly, and it is clear that they could not underwrite any expensive new public-works projects such as draining swamps or constructing pumps. Nevertheless, as judged by their actions, something more complex than economic constraints prompted their unwillingness to work without funding. Their rhetoric overstated their position.

Two patterns in their behavior underscore this point. First, under Huet's "patronage" they paid the expenses of their individual projects and expected to contribute to the academy's general operating expenses by furnishing such things as dissection specimens and the materials needed to construct new instruments. Indeed, even as they refused to work in Chamillart's name, they continued to bear the expenses of the program they had followed under Huet. Second, although they made their case in terms of the vast expenses involved in draining swamps and building pumps, they proved just as unwilling to undertake projects involving little or no expense. Undoubtedly, had Huet asked, the academicians would have put a weathervane on his house; for Chamillart they would do nothing.

As for Graindorge, funding was something he thought Chamillart should provide, but for him that issue paled in significance next to what he really wanted from this new patron/*chef*. Graindorge willingly paid for his own research.[19] Moreover, on at least two occasions during the first months under Chamillart he offered to finance projects for other academicians. Of course, coming from Graindorge, such an offer could become counterproductive, as it was when he told

[19]See, for example, Graindorge's letter of 30 December 1667: "Il y a un petit traitté du nitre qui est le plus joly du monde et le plus aisé a pratiquer que nous executerons quand vous serez icy a peu de frais. En cas que nous nayons pas des fonds nous boursillerons vous et moy as dessus en notre particuliers avec le Sr Hauton" (BL, A 1866, 570).

Lasson he would help finance the construction of a new burning mirror.[20] Graindorge did not possess the *état* necessary to become Lasson's patron, and the insult led Lasson to stop work altogether. For Graindorge, ample financing was comparable to an exciting *curiosité*—it was important, but only if (and when) the academicians came to the laboratory.

Even as the academy accepted Chamillart's "patronage," Huet continued to underwrite the academy's major financial burden. According to Graindorge, these costs were minimal. Huet gave the organization the use of his house, its furnishings, and the instruments he owned. Beyond those overhead costs, Huet's only regular out-of-pocket expenses involved the candles and firewood consumed during sessions. He even expected the academicians to share the costs of anatomy specimens.[21] For Chamillart to become an effective patron he had to give some other sign of performing a patronage function. The best thing he could have done was just what Graindorge wanted him to do—simply come to the sessions. With Huet behind the scenes and Chamillart listening to the academicians produce their science each week, Graindorge knew he could make the academy work. As the second session under the new statutes demonstrates, he was probably right.

Everyone expected Chamillart to come to the academy on 10 November, but like Huet, he was out of town. Unlike Huet, however, Chamillart failed to tell anyone he would not attend. The expectation of working for the intendant produced a full turnout of the membership (minus Huet, of course). Thus, by accident, Graindorge had a chance to direct a session attended by the full complement of the available membership. It may well have been the first such session. At first things did not go too badly. The newly enrolled academicians Chasles presented a demonstration of surveying techniques that held everyone's interest. When Graindorge turned to the dissection he had prepared, however, the session fell apart: "I had brought two small dogs to use as anatomy specimens. . . . Our mathematicians, who hardly care for anatomy, and who care even less for lymphatics, got up and left us there, so that we stopped before the appointed hour."[22] Graindorge excused this rudeness, saying the hour was late, the weather was terrible, and the light was bad. Nevertheless, the point of

[20]Ibid., 571 (13 January 1668).
[21]Ibid., 597 (27 January 1670).
[22]Ibid., 564 (11 November 1667).

what he reported is all too clear. Without Chamillart or Huet there, at least half the academicians left when Graindorge started dissecting. He lacked the power to hold them in the laboratory.

Nonattendance was not unique to this academy, nor was it peculiarly directed at Graindorge. Getting academicians to their sessions was a problem Colbert faced even at the most prestigious royal academies in Paris. Both the Cimento in Florence and the Royal Society of London faced similar difficulties.[23] In Paris, nonattendance proved particularly disruptive to the Académie Française and the Académie Royale des Sciences. Instead of physically seating himself at each session of those academies, Colbert instituted what became known as the *jeton* system, under which the academy's secretary distributed tokens redeemable for cash at each session.[24] Colbert found that these *jetons de présence* were the only means to guarantee that the Parisian academicians would attend their sessions—or if they came, that they would stay for the entire time required.[25] Obviously the *jeton* system was unfeasible in Caen;[26] but then, Graindorge thought the academy could rely on Chamillart's presence to accomplish the same end. He was correct in his assessment of the effect Chamillart's attendance would have; he was mistaken only about how often Chamillart was going to appear at sessions.

Chamillart's position in Caen gave him the power to animate the

[23]For an extended discussion on the sporadic nature of the Cimento's sessions, see Middleton, *Experimenters,* pp. 61–64. For a similar discussion of problems with both membership and attendance at the Royal Society, see Michael Hunter, *Science and Society in Restoration England* (Cambridge, 1981), pp. 42–46.

[24]*LIMC,* vol. 5, *Compliment de Charpentier,* pp. 542–543. This document explains the *jeton* system as it applied to the Académie Française. The monetary consideration involved was considerable. For example, Pierre-Daniel Huet's account book (BN, Fr, n.a., 1197, f. 269v) reveals that between July 1675 and November 1676 the value of the *jetons* was just over 400 livres. For an account of how and why Colbert instituted the system, see Charles Perrault, *Mémoires de ma vie,* ed. Paul Bonnefon (Paris, 1909), pp. 97–98.

[25]Under the *jeton* system used at the Académie Française, the secretary made a list of the members present when the session opened. Then he drew up a second list at the session's end. Only those academicians whose names appeared on both lists shared in the distribution of *jetons.* The number of *jetons* an academician received depended on attendance at each session, as the secretary made an equal division of the available "pot" among all those present. Any tokens remaining after that equal division were returned to the "pot" and added to the regular allotment furnished for the next session.

[26]According to Paul Bonnefon, the editor of Perrault's *Mémoires,* the *jeton* system at the Académie Française cost 6,400 livres a year when Colbert instituted it. It was the major expense for that academy, which operated on an annual budget of only 7,000 livres (Perrault, *Mémoires,* p. 98).

academy, but before we can go any further with questions about his impact on the academicians, we must consider what the intendant wanted from them, as well as what he thought he was giving in return. Determination of what he sought from the academy is a fairly simple task. If one considers what Graindorge reported together with what is known more generally about Chamillart's intendance, his desires become clear. He sought technical help with the administration of his generality, and he thought the academy could assist him in two areas.

Beginning with the extraordinary taxation measures and the consequent revolts of the 1630s, Lower Normandy had entered a deep economic slump.[27] Chamillart's intendance marked the beginning of a revitalization. He worked to establish new industries, especially textiles, and undertook special measures to promote trade. As intendant, he was thus an activist promoting the economic development of his generality. If the academicians helped him with technical assistance, they could contribute to the welfare of their own province.

Moreover, Chamillart clearly recognized his more direct responsibilities as the king's agent in Normandy. He wanted to maximize the economic utility of all royal rights that fell under his jurisdiction. As one of Colbert's *fidèles,* he had already worked in that minister's administration of the royal *eaux et forêts,* and as intendant in Lower Normandy he wanted to exploit all those natural resources that belonged to the king, especially mineral and timber rights. In addition to more general economic development, then, he particularly wanted to extract resources from the royal domain (mining and the production of timber for naval stores), an activity in which the academy could render direct service to the monarchy.

As we shall see repeatedly, this general portrait of Chamillart as a "progressive" royal official extracting natural resources from royal lands while also working for the economic revitalization of his generality finds its direct parallel in the projects he urged on the Académie de Physique. We have already noted his desire that the academy undertake two projects in hydraulic engineering—the draining of swamps and the construction of pumps for fountains. At other times he asked for projects that expressed his desire to promote new economic activity and exploit royal rights. At various times he wanted an estimate of the timber a royal forest could produce, an improved technique for cannon boring, development of new dyes from shell-

[27]For discussion, see A. Galland, *L'Histoire du Protestantisme à Caen et en Basse-Normandie de l'Édit de Nantes à la Révolution* (Paris, 1898), pp. 104–109.

fish, a method for clearing a channel that would make the river Orne navigable as far upstream as Caen, a survey of the generality's available deposits of lead, and a study of the efficiency of various hull designs.

That list of projects clearly indicates that he saw the academy as an adjunct technical bureau of his intendance. Chamillart was not interested in promoting research in Graindorge's empirical science, nor was he in the least interested in gaining glory as a patron in the Republic of Letters. He wanted a technical research institute that could solve practical problems and fabricate real devices. According to Graindorge, "such things are more important to the intendant than all our reasoning and speculations."[28]

Chamillart wanted something other than pure intellectual activity from the academy, that much is clear. What he thought he gave in return is more difficult to define. He had to have something to offer or he could never hope to wean Huet's academy from its nominal (if at times unenthusiastic) pursuit of *curiosités* and *expériences*. Here we must be cautious not only because we lack Chamillart's direct testimony on the subject but also because Graindorge so obviously misunderstood what the intendant was doing.[29] Nevertheless, Graindorge's descriptions of Chamillart's behavior offer a consistent picture. Despite Chamillart's obvious lack of interest in securing letters patent for the academy, it is clear that he felt his most important contribution as *chef* was in his ability to offer the organization status and legitimacy within the corporate structure of contemporary France: he was offering the legal fiction of a corporate personality. In fact, this was a valuable commodity in seventeenth-century France.

Before 1667, however, there is little evidence to suggest that the academy actually needed corporate legitimation. Only once do Graindorge's letters hint that Huet's Académie de Physique had ever attracted critical notice from any more established corporate interests. That had come in the fall and winter of 1665, when Huet was first beginning to hold the regular weekly sessions that stamped his group with the term "academy." Moisant de Brieux claimed that his Grand Cheval already possessed corporate standing,[30] and twice during the time Huet was regularizing his group Graindorge made remarks that

[28]BL, A 1866, 565 (18 November 1667).

[29]On this subject, in fact, Graindorge's direct statements to Huet were often unreliable and contradictory. In effect, they amounted to personal statements on what he thought the intendant *should do*.

[30]For Moisant de Brieux's own statement on this subject in 1669, see François-Richard de La Londe, *Mémoire pour servir à l'historie de l'histoire de l'académie des belles-*

suggest Brieux was questioning Huet's right to form any new "academy."[31] Ultimately this problem, if it ever truly existed, turned out to be insignificant. The Académie de Physique certainly did not need Chamillart's protection from the Grand Cheval.

Nevertheless, the issue was real. A private-patronage academy, especially one that dealt with natural philosophy, was at least technically vulnerable to a variety of accusations concerning illegal or immoral behavior. Given the restrictions the Sorbonne and the corporate interests in Paris had just imposed on the Académie Royale des Sciences (during 1666–1667),[32] Huet's academy worked under the very real possibility of corporate interference with its activities. Hauton's alchemy, Huet's blood transfusions, Busnel's human dissections, too great an interest in the astrological aspects of astronomy—any such activities could have led to a legal entanglement with the university in Caen if the academy became too vocal in trumpeting its scientific successes. In the fall of 1667, in fact, at least some at the university decided that the academy had become a problem.

During the last week of October, Graindorge heard rumors that the Faculty of Medicine planned to prohibit the surgeon Busnel from doing anatomy demonstrations for the Académie de Physique.[33] When he first mentioned that possibility to Huet, Graindorge had no details on the prohibition or the reasons for it. He soon learned, however, that the Faculty of Medicine had planned the prohibition when it learned that the Académie de Physique might become a royal institution. If that was the case, the organization had become a threat to the university:

> Busnel has not heard any talk about this prohibition. Nevertheless, there are several things to say on the topic. In a word, M. de Vaucouleurs, with his own ears, heard M. Verel say that all the academicians should be expelled from the university. . . . They say that the more privileges there are, the less they enjoy theirs, and that [our academy] will divert their wages or prevent them from being increased. I was told even that they believe [our academy] is intended to establish Cartesianism.[34]

lettres de Caen (Caen, 1854), pp. 9–11. For a more extensive discussion, see Brennan, "Culture and Dependencies," pp. 75–80.

[31]These letters are discussed in chap. 2 above, under "The Mechanics of Change."

[32]See George, "Genesis," p. 386; Brown, *Scientific Organizations*, pp. 147–148. For an explanation of how the restrictions demanded by the Sorbonne fitted Colbert's overall plans for "la Grande Académie," see Perrault, *Mémoires de ma vie*, pp. 42–48.

[33]BL, A 1866, 675 (31 October 1667).

[34]Ibid., 564 (11 November 1667).

The threat to the academy's existence was real. Actually it had existed since Huet and Graindorge first began to host the early *assemblée.* In a corporate society, only letters patent or the protection of a powerful patron such as Chamillart could protect the academy from the possibility of a legal entanglement with the university.

Of course, there is an intriguing and ironic side to the notion of Chamillart protecting the academy from the kind of opposition M. Verel voiced. Undoubtedly the academy needed a powerful patron/protector if it hoped to survive for long while pursuing a scientific program such as that envisioned by Graindorge. But the academy that existed in late 1667 was not yet the one he wanted. Clearly Verel's fears were a response to the threat posed by a new royal corporation. Verel described those fears in relation to privilege, wages, and Cartesianism—specific threats only a royal academy could pose. Huet's private-patronage society had yet to trouble anyone. A royal academy of sciences, however, was an entirely different matter. From the university's perspective, the mere existence of such an institution threatened its social, economic, and intellectual privileges. The academy needed corporate legitimacy to succeed, but ironically, Chamillart himself created the organization's first real need for such protection—just by having the power to confer protection.[35]

In the coinage of patronage, corporate legitimation was valuable. But to claim it was genuinely meaningful to the academy in late 1667 would be foolish. Like a viable intellectual program and adequate financial resources, corporate legitimation was something a patron might provide, but for the Académie de Physique it was a low priority. In late 1667 the organization needed a *chef* who could direct its week-to-week operation. That was a role Chamillart was unwilling to assume. He showed no inclination to become a working member of the academy.

Following Chamillart's first appearance at the academy, he did not come again for six weeks. Week by week over that time Graindorge's letters chronicled the disintegration of the group. We have already seen what happened at the session on 10 November: half the members left when Chamillart failed to appear. The following week

[35]This issue affected the academicians' thinking. In January 1668 Graindorge told Huet that both Savary and Vaucouleurs were planning to publish some of their literary works and that Savary intended to claim he was "dAcademia regia socius," whereas Vaucouleurs did not: "a cause de luiniversité dont il est membre, il na pas voulu y adiouster cette qualité" (ibid., 572 [16 January 1668]).

(17 November) poor attendance "crippled" the session and the "talk was vague." The next week only two academicians appeared at Huet's house, and Graindorge concluded that "our academy is being robbed little by little." On 1 December more academicians attended, but Graindorge claimed that things were starting to happen only "bit by bit." A week later (8 December) Graindorge canceled the session because of a local holiday. By that time, even as Graindorge reported some good news (that the intendant was planning to come to the next session), he was beginning to hope for Huet's prompt return "to animate our academy."[36]

After his first session as *chef* (3 November), Chamillart did not attend another until 15 December, the seventh session nominally under his direction. During the time between these appearances, attendance had fallen off dramatically and the sessions had come and gone with little activity beyond some dissections, a few technical demonstrations, and some vague planning for building a celestial globe.[37] The academy was dead in the water. All that changed, however, as soon as the intendant announced he would come on 15 December. With the intendant's reappearance, even Graindorge lost his recent cynicism:

> You will have heard that the intendant came to our academy, which he animated with a speech full of good intentions and grand projects. No one could talk about anything but making rivers navigable, constructing new seaports, equipping the navy, building machines, improving navigation, and illustrating astronomy—in a word, the greatest possible plans. . . .
>
> The intendant took our names to send to court. Finally he promised that he will request something for us as a New Year's gift. If this happens, hurry your return because you will see us in action up to our ears.[38]

Graindorge had been correct: the most important thing the academy needed from Chamillart was nothing more than his presence and some words of direct encouragement, especially when those words promised a royal "New Year's gift."

With his fine speech, Chamillart extracted a long list of proposals for new projects from the academicians. Moreover, he had pledges

[36] Ibid., 565, 566, 654, 568 (18 and 23 November, 4 and 12 December 1667).
[37] Ibid., 564–567, 654, 568.
[38] Ibid., 569 (19 December 1667).

from several members that they would begin this work on their own initiative. Graindorge, for example, promised to undertake three specific tasks immediately, and Lasson accepted commissions to map the river Orne, inspect a quarry, and examine all the swamps in the generality. None of these projects involved great expense, but that was not the issue. What put the men to work (or at least prompted them to promise work) was the fact that Chamillart finally seemed to act like a patron/*chef.*

When the session held on 15 December 1667 adjourned, the academy gave every appearance of being something completely different from what it had ever been before. Of course, one successful session spent drawing up plans was not likely to transform the Académie de Physique. Within a week this newfound enthusiasm dissipated, for some obvious reasons. As he reported this session to Huet, however, Graindorge was euphoric. He must have savored the moment, and before continuing with our account of the academy's incorporation, we should ask why. To find an answer we need to look more closely both at the men who formed Chamillart's academy and at the program they pledged to deliver.

CHAMILLART'S ACADEMY

When the academicians gave their response to Chamillart's proposals at the 15 December session, Graindorge once again became excited about the academy's future. He had good reason for his optimism. The program Chamillart sketched out was ambitious, but many of the projects (at least those with specific focus) could be done. Indeed, over the next five years the academicians did make progress in working on a fair number of these projects. The academy boasted a membership that offered a great deal of expertise. Graindorge never saw the academy's problem as a lack of talent; the problem was always in marshaling that talent for some useful purpose.

The New Men

Besides the six men who had belonged to the academy in 1666—Huet, Graindorge, Villons, Lasson, Vaucouleurs, and Hauton—three more had been taken on as full members before Chamillart had become the *chef.* Chamillart had named one new academician (at

Graindorge's suggestion) during their negotiations over the statutes.[39] Thus at the time of Chamillart's takeover, the academy possessed its full complement of ten members. In principle, the four new members added during 1667 broke down evenly as two new *mathématiciens* and two new *physiciens,* thus maintaining an exact balance between the academy's two sections. In practice, things did not work out so neatly, but the classification given in the statutes was not far from reality.

The man whose presence in the academy actually skewed the balance between *mathématiciens* and *physiciens* was Jacques Savary, sieur de Courtsigny.[40] Savary was the last of the ten academicians named to hold a chair, and he was in fact the only one Chamillart actually had any voice in selecting. According to the way Graindorge listed Savary in the statutes, he was a *mathématicien.* At first this classification is puzzling. Indeed, his selection as an academician is odd, especially since at least three other men had also sought the chair he filled.[41] Graindorge's letters indicate that he had never had any association with the group until Chamillart was preparing to take over, and that he received his chair over the objections of some of the other academicians. Although Graindorge's letters never specified what those objections were, they undoubtedly related to the fact that Savary had no obvious qualifications as either a *mathématicien* or a *physicien.* He was known locally as an avid litterateur, poet, and huntsman.[42]

The mystery over Savary's qualifications as an academician resolves itself when we find he read English. He was not only the one member of the academy who possessed this specialized skill but one of the few in Caen who could undertake a task as complex as translating the *Philosophical Transactions* into French. Apparently his work went slowly, but he did produce translations faster than they could be

[39]Ibid., 564 (11 November 1667).

[40]For biographical information on Savary, see Tolmer, *Huet,* pp. 356–357.

[41]Graindorge admitted two of these men, Pierre le Vavasseur and Nicolas Postel, to the academy as supernumeraries during November 1667. The third applicant for the academy's tenth chair was Antoine Halley, Huet's former teacher. Halley's letter applying for this chair has survived among Huet's papers (BL, A 1866, 723). Although L. G. Pellissier dated the letter to 1662 on the basis of its description of a "new royal academy," a careful reading makes it clear that Halley's letter must have been written during the fall of 1667. Halley's failure to appear in the academy is intriguing. For biographical information, see Victor Evremont Pillet, "Étude sur Antoine Halley," *MANC,* 1858, pp. 173–224.

[42]For Huet's evaluation of Savary's intellectual capabilities, see *Commentarius,* pp. 158–159.

borrowed, bought, or commissioned in Paris. Huet and Graindorge agreed that access to the translations was essential to the academy if it was to become truly productive in chasing down *curiosités.*[43] In fact, before Savary joined the academy, Huet carried on an extensive correspondence with Henri Justel in an effort to find some way to get translations from Paris. Savary was thus a valuable resource, but of course his value depended on the existence of an active research program based on *curiosités.* Until Graindorge established such a program, Savary would hold one of the academy's ten chairs but make no effective contribution to the research program.

In contrast to Savary, the other man named as a new *mathématicien* proved to be one of the most devoted of the academy's members. Jacques Chasles in fact proved to be the only one of the five *mathématiciens* to follow through on all the work he promised.[44] He also attended sessions regularly, produced demonstrations, and eagerly undertook new assignments from Graindorge. In January 1668, for instance, he began the laborious (and ultimately thankless) task of drawing up conversion tables for the various systems of weights and measures in common use throughout France. He was still at this task five years later. Unfortunately, Chasles was something less than gifted. Even Graindorge, who had every reason to be grateful for the man's faithfulness to the academy, finally had to admit, "Chasles has good sense, but he is slow."[45] Undoubtedly, as the plans for the new program were being made, Chasles was considered the least important of the *mathématiciens.* The others—Huet, Villons, Lasson, and Savary—gave the appearance of great potential, but Chasles's dogged persistence actually made him the only academician who worked regularly from 1667 through 1672 as a *mathématicien.*

With the two new *physiciens,* the situation was much the same—significant potential, equivocal performance. The most promising of the new additions was Pierre Cally, the only one of the five *physiciens*

[43]During late 1667 and early 1668 Graindorge's letters frequently refer to his eagerness to have access to the latest issues of the *Transactions.* In his letter of 13 January 1668, for example, Graindorge told Huet: "Mr Savary qui receu hier matin assez tard la transaction d'angleterre nous apporta pres de deux pages d'interpretées." Since "yesterday" for this letter was a Thursday, Savary had translated his two pages during the afternoon in order to have them ready for the evening's session (BL, A 1866, 571). For a general discussion of French enthusiasm for the *Philosophical Transactions,* see Brown, *Scientific Organizations,* pp. 200–207.

[44]Little biographical information on Chasles is available. For discussion, see Tolmer, *Huet,* pp. 357–358.

[45]BL, A 1866, 601 (21 April 1667).

who was not a *médecin.* Cally was a professor of philosophy at the Collège du Bois in Caen and was known as both an expert cartographer and an outspoken Cartesian philosopher.[46] Possibly Cally's presence among the academicians prompted the university's fear of the academy as a center of Cartesianism. They need not have worried. Like Huet, Lasson, Villons, and Savary, Cally attended sessions only sporadically. In fact, his name appears less often in Graindorge's letters to Huet than any other academician's. After several years of dealing with Cally, the most charitable comment Graindorge could find to make on his contribution to the group was "Cally listens."[47]

The other new *physicien* named in the statutes was Jean-Baptiste Callard de La Ducquerie.[48] Like all the others in his section except Cally, he was a *médecin.* Moreover, like Vaucouleurs, he was both a traditional Galenist and a professor of medicine at the university. Whereas Vaucouleurs's special expertise lay in anatomy and physiology, La Ducquerie specialized in Galenic pharmacology. Undoubtedly his presence among the membership is the reason the academy's statutes specified that the group would maintain an herb garden. Unfortunately, in late 1667 the estimated costs of such a garden exceeded 1,000 livres. Once that fact became known, Chamillart's position on new funding meant that La Ducquerie actually made his contribution to the academy by joining Graindorge and Vaucouleurs at the dissecting table. Not surprisingly, he often proved less than enthusiastic about such work, and Graindorge finally had to conclude: "La Ducquerie only flaps one wing."[49]

In addition to the ten academicians, Chamillart, and Busnel—all of whom are listed in the statutes as filling their respective roles—Graindorge added two supernumerary members to the group during November 1667.[50] Apparently he did so on his own authority. Most of the other academicians avoided the sessions when Huet and Chamillart were unavailable, but these men were eager to attend. One was Pierre Le Vavasseur, a surveyor whom Graindorge expected to help fulfill Chamillart's desire to map the generality.[51] The other was

[46]Tolmer, *Huet,* pp. 358–366.

[47]BL, A 1866, 601 (21 April 1670).

[48]Tolmer, *Huet,* pp. 354–355.

[49]BL, A 1866, 601 (21 April 1670).

[50]Graindorge definitely considered both these additions "extra" academicians, as is demonstrated by his admission to Huet that he had not sought permission from Chamillart to admit them to sessions (ibid., 654 [4 December 1667]).

[51]Tolmer, *Huet,* p. 368.

Nicolas Postel, a professor of medicine at the university at Caen who was to assist La Ducquerie with the academy's herb garden.[52] Graindorge made no apology for admitting these men to sessions, saying (in Vavasseur's case) it was "because one finds so few who possess the diligence [the academy] so often lacks."[53] Apparently Vavasseur was regular in his attendance, but like the *mathématicien* Chasles, he was a plodder. Graindorge finally had to admit to Huet that Vavasseur was "no great operator." Postel's name appears only infrequently in Graindorge's letters, and it seems he was never very active in the academy's work. After all, there was no money for an herb garden. After more than two years of dealing with Postel, Graindorge finally concluded that he had yet to make a single "reasonable" contribution to the academy.[54]

Chamillart's Program

The list of projects Graindorge and the academicians promised to undertake for Chamillart is too long and cumbersome to quote here.[55] Moreover, as Graindorge himself told Huet, some of the projects were "superfluous," either because they were too vague (for example, "perfect the [technical] arts"; "research with care knowledge that will serve the health, utility, and convenience of life") or because they required expertise that was clearly not available among the academicians (design new ships).[56] A much better approach to explaining what the academicians promised at their 15 December session is simply to use that list as a basis for dividing their projects into four broad areas, or subprograms.

The first and most prominently featured of these subprograms was in astronomy.[57] Several academicians had roles to play in this area. An abandoned tower just outside the city could make a wonderful observatory. Since the tower of Chatimoine belonged to the city, and since Graindorge was an *échevin de ville,* or magistrate, he was the natural person to secure its use as an observatory, and he did so almost immediately. Next, to equip an up-to-date astronomical obser-

[52]Brown, "L'Académie de Physique," p. 191.
[53]BL, A 1866, 565 (18 November 1667).
[54]Ibid., 601 (21 April 1670).
[55]Brown has published the text of these resolutions in his "Académie de Physique," pp. 155–156.
[56]BL, A 1866, 570 (30 December 1667).
[57]Ibid.

vatory for the 1660s, the academy needed telescopes and pendulum clocks. The expertise for constructing these devices clearly belonged to Lasson (telescopes) and Villons (clocks). During this period Lasson was already promising a parabolic lens (he never produced one) and Villons was about to design a new clock escapement mechanism that would bring him considerable attention in Paris—although he never actually perfected the device. For observers, the academy had any number of candidates, with Huet leading the list of those most interested in this activity. Given what we already know about the personnel involved, it should come as no surprise to learn that the program in observational astronomy never got under way. Graindorge secured the use of the tower of Chatimoine, but that was as much progress as the academy ever made with this part of the astronomy program.

Nevertheless, as with each of the other three subprograms proposed to Chamillart, we should not rush to judgment on the success or failure of the academy's astronomy program. In addition to securing the tower of Chatimoine for the academy's use, Graindorge had also promised to coax the secret of a solution to the longitude problem from a reclusive Norman savant who for years had claimed he would reveal his work to no one but the king of France himself. The savant, Jacques Graindorge, was the prior of a Benedictine abbey in Fontenay, near Caen. He was also a cousin of André Graindorge.[58]

As early as 1662 Huet had known that Jacques Graindorge claimed to possess the secret of longitudes. He also knew that Jacques Graindorge claimed to be able to observe the motions of stars during daylight hours, predict shifts in the winds, and offer a new theory to explain the tides. The monk was far from shy about trumpeting his accomplishments, but he refused to tell anyone how he did these things. He had even written to the king's minister Colbert to proclaim his expertise; the intelligencer Henri Justel had heard about his claims from correspondents as far away as the Netherlands.[59] The monk's claims had created high expectations far beyond the limits of Caen. He claimed to have solved the longitude problem by determining a "fixed meridian" on the earth's surface, which in turn was determined by the earth's position against the fixed stars. In other words, he claimed to enable mariners to determine lines of longitude by direct observation, just as they determined latitude on the earth's

[58]Graindorge's discussions of his difficulties with his cousin first appear in January 1668 (ibid., 703, 571, 573 [6, 13, and 27 January 1668]).

[59]Tolmer, *Huet*, p. 310.

surface. All in all, Graindorge had little trouble convincing Chamillart that the academy should include verification of the monk's extravagant claims among the *curiosités* to be investigated in the new program.

Graindorge faced innumerable frustrations in his dealings with the "surly monk" and never did persuade him to reveal his secret before the membership of the Académie de Physique. Nonetheless, he and Chamillart finally persuaded his cousin to present the method directly to the Académie Royale. In early 1669 the monk made the trip to Paris for that presentation with the Académie de Physique's recommendation, and with Colbert paying his expenses. The Académie Royale received the monk graciously and then detailed a committee composed of Christiaan Huygens and Jean Picard to study the method. Although these professionals finally declared his method fallacious ("He has provided nothing that he promised"),[60] the episode redounded to the Académie de Physique's credit.

Just as Graindorge claimed, in the 1660s the body of knowledge that constituted natural philosophy contained innumerable plausible but unsupported claims put forward by various savants and *curieux.*[61] Likewise, as he argued, *expériences* and organized checking of the validity of such claims offered a powerful antidote to extravagant and poorly founded claims such as those of Jacques Graindorge. Actually, by the time the monk was ready to deliver his ideas in Paris, neither Graindorge nor Huet believed that there was anything of substance to his claims.[62] Yet, following the programmatic ideas he had developed while attending the Thévenot, Graindorge could not allow himself to dismiss such claims out of hand. Since his adoption of empiricism at the Thévenot, he had found that many long-held ideas were fallacious, whereas apparently absurd claims sometimes proved true.

The academy's dealings with the monk were as successful for Graindorge's scientific program as the experiments with oil of tobacco, the barometer, and transfusions. Moreover, in exposing this particular *curiosité* as groundless, the academy had rendered a real service to the monarchy. Given the stakes involved in finding an acceptable method for determining longitudes at sea, Colbert could not ignore any proposal that seemed to offer even minimal prospects

[60]AdS, Reg (Mathématiques), vol. 3, ff. 261r–276v.

[61]For an excellent discussion of this issue in contemporary English science, see Hunter, *Science and Society in Restoration England,* pp. 87–112.

[62]Tolmer, *Huet,* pp. 310–313.

for success. Before his presentation in Paris, Jacques Graindorge's method appeared to offer more than that minimal chance. It had become a part of regional lore in Normandy, and the idea was being pursued in both London and the Netherlands; but even more important, the academicians in Paris were unable to deny its validity until they had a chance to study the proposal in detail. In sponsoring this submission to the Académie Royale, then, the Académie de Physique helped to eliminate not only the specific *curiosité* Jacques Graindorge presented but all generic proposals that claimed to solve the longitude problem through the determination of a fixed meridian. The entire episode actually validated Graindorge's assertion that organized scientific societies provided the means to confirm or deny unverified claims about nature.

In the academy's second subprogram the membership also had marked success—although it still failed to fulfill the expectations created in its resolutions of 15 December 1667.[63] These tasks are best placed under the general heading of civil engineering, but to the academicians of the 1660s they represented a variety of specializations: constructing pumps, producing maps and coastal charts, draining swamps, clearing a navigable channel in the river Orne, improving fortifications and constructing ports, and perfecting a device for the desalinization of seawater. Of the projects on that list, the ones that produced the most tangible results were the desalinization of seawater (for which the academician Hauton would in fact receive a royal *gratification*) and a study of the feasibility of clearing a channel in the Orne.[64] As for the rest, Cally's expertise as a cartographer produced nothing; Chasles and Vavasseur never put their surveying skills to use except in demonstrations before the academy; and Lasson and Villons never put their skills with machinery to work draining swamps or building pumps.

Hauton's seawater project, however, offers one of the best examples of the academy's research capabilities. At the time of the royal incorporation in late 1667, Hauton was already at work on this project. He had probably been working on it since mid-1667;[65] if he had, the coming of the royal incorporation caused him to redouble his

[63]BL, A 1866, 570 (30 December 1667).

[64]Villons concluded it was impossible, but did complete a new chart of the river (ibid., 591, 683, 708 (26 July 1669 and [mid-1669]).

[65]Ibid., 562 (4 July 1667). Graindorge suggests several distillation projects to Huet, specifically suggesting that either Hauton or La Ducquerie might pursue them.

efforts. The problem he had identified sounded simple, yet the project's execution entailed more than two years of sustained effort. Hauton's desalinization project required him to develop the apparatus, techniques, and materials necessary to supply oceangoing ships with emergency drinking water.[66] By any standard, his work constituted a sophisticated technological achievement. Over the course of the two years Hauton worked on the project, he not only set himself to design an apparatus to combine precipitation, distillation, and filtration, but also dealt with such problems as reducing the size of his equipment to fit the limited space available on contemporary ships; balancing the cost of operation against the system's daily output (approximately eighty gallons at a cost of ten sous); meeting requirements for the water's taste; and ensuring reliable, continuous operation on board busy, crowded ships. This project went far beyond the creation of a new device. It produced, in fact, a complete system to furnish emergency drinking water. When Hauton found, for example, that the materials he used in his filtration process threw his cost projections out of line, he devised a new process for producing the filter medium more cheaply. Likewise, to save space, speed the system's operation, and reduce the manpower required for operation, he simplified the condenser's design to eliminate the need to replenish the water supply in the cooling jacket of the heat exchanger.[67]

Hauton certainly earned the *approbation* that the Académie Royale gave his work following a demonstration in late 1669. He also earned the 1,200 livres that Colbert granted in the king's name. This was exactly the kind of project Chamillart and Colbert wanted from the Académie de Physique. It was a practical solution to a real problem.

The same kind of practicality characterized another of the academy's projects, one that belonged to the civil engineering program but did not gain a specific reference in the resolutions of 15 December—the academician Chasles's attempt to devise a rationalized system of weights and measures for the kingdom of France. This project did not appear in the resolutions offered to Chamillart because Chasles had not yet formulated the research. By the first week of

[66]Ibid., 595 (20 December 1669).

[67]He piped his apparatus so as to run the condensation coil through the hull of the ship for direct cooling in the ocean. The best available description of Hauton's device appears in Huet's account for Henry Oldenburg (12 February 1669/70, *CHO,* 6:486–488).

1668, however, the issue of weights and measures had come before the academy, and Chasles promptly set to work on his problem. Thus, despite its omission from the resolutions of 15 December, the project was clearly a part of the reforms Chamillart brought to the academy.

Originally the project was a simple empirical exercise: Chasles tried to establish conversion tables. Starting from the commonly accepted notion that the chaotic state of French weights and measures in the 1660s resulted from a lack of political will to carry through a needed reform, Chasles set out to create a simple conversion formula that the monarchy could impose throughout France. The effort ran into trouble on two fronts. The first was the difficulty inherent in an attempt to gather accurate exemplars from the multitude of local guilds scattered in cities throughout France, each of which maintained its own private bureau of standards. The second came as Chasles finally obtained enough exemplars from local systems of weights and measures to realize that direct commensurability was a chimera. No conceivable standard could possibly serve to convert the myriad disparate measures into rational forms. He took the next logical step of seeking a single, naturally occurring unit that could serve as the basis of a new, universal system of weights and measures.[68] In doing so, he also conceived the desirability of designing a system that would tie together measurements of weight, volume, and linear dimensions.

The difficulty of evaluating Chasles's project comes when one tries to place it within the scientific and technological context of the 1660s. Obviously he was pursuing a fruitful line of inquiry, since the formulation of the metric system more than a century later followed exactly the same rationale. Just as clearly, however, this one man had little chance of reforming French weights and measures. In the 1660s Chasles was doomed to fail. Nevertheless it was an important project. Simply by going through the tedious attempt to create conversion tables, Chasles demonstrated the futility of this approach to the rationalization of French weights and measures. Then, in attempting to find a new universal standard in nature, he demonstrated the difficulties inherent in that approach. In effect, his achievement was negative but substantial. In failing to solve his problem, he destroyed the assumptions that had prompted him to undertake the project in the first place: he showed conclusively that something more than the will to reform was needed. Moreover, he defined the rationalization of weights and measures through the identification of naturally oc-

[68]BL, A 1866, 703, 583 (6 and 25 January 1669), 604, 655, and 656.

curring units as a substantial conceptual and technical problem, anticipating both Picard and Huygens by several years.[69]

In the third general area in which the academicians promised to work—metallurgy and the fabrication of scientific instruments—they produced consequential results, even if once again they failed to fulfill their resolutions. By necessity, any number of academicians built devices in the course of producing demonstrations for the group. The anatomist Vaucouleurs had built his own barometer, for example; Chasles and La Ducquerie produced numerous interesting devices and instruments; even Lasson and Villons presented some of their handiwork on occasion. Nevertheless, this level of metallurgy and instrument making fell far short of what the academicians had promised to Chamillart: a celestial globe five feet in diameter (Villons), the largest burning mirror ever cast (Lasson), new pendulum clocks (Villons), microscopes (Lasson?), improved techniques for boring cannon and constructing muskets (Villons), and the fabrication of basic laboratory instruments and construction of a chemical furnace. With the exception of Lasson's burning mirror and Villons's radically new clock mechanism, none of these projects ever got beyond the planning stages. And except for Villons's marine chronometer (his new clock design), the academy's ambitious-sounding program in metallurgy and instrument making never produced anything that could be presented in Paris.

The task of fabricating the largest burning mirror ever cast was entirely within Lasson's technical capabilities. Indeed, at the time of the royal incorporation he had already built the mold—he claimed he lacked only the bell metal necessary for the casting.[70] More accurately, perhaps, he lacked the 150 to 200 livres that he estimated the metal would cost. Such a sum was not inconceivable, but according to the kind of figures recorded in Huet's daily account books, it could have supported Huet comfortably in Paris for about two months, in Rouen for six months.[71] In other words, it was a sum large enough to

[69]For discussions of these problems as they appeared in the eighteenth century, see Keith Michael Baker, *Condorcet: From Natural Philosophy to Social Mathematics* (Chicago, 1975), pp. 65–67; Maurice Crosland, "'Nature' and Measurement in Eighteenth-Century France," *Studies on Voltaire and the Eighteenth Century* 87 (1972): 277–309. For an intriguing and suggestive account of the conceptual difficulties seventeenth-century scientists could encounter in dealing with weights and measures in the laboratory, see T. S. Patterson, "Van Helmont's Ice and Water Experiments," *AS* 1 (1936): 462–467.

[70]BL, A 1866, 564, 703, 571 (11 November 1667, 6 and 13 January 1668).

[71]BN, Fr, n.a., 1197.

cool Lasson's enthusiasm for financing the project himself. Chamillart refused to provide the money, and when Graindorge offered to buy the necessary materials, Lasson was so insulted he refused to continue with the project. More than eighteen months later, Lasson still had the mold ready but steadfastly refused to complete the mirror.[72] In short, the project was destroyed by Chamillart and Colbert's funding policy.

Such was not the case with Villons's efforts at clockmaking. Indeed, as Chamillart and Colbert could well have claimed, the rewards Villons received from his investment in constructing a new form of escapement were what they had in mind for any academician who demonstrated his worth. Villons engineered a genuine technological breakthrough even though his clock never proved trustworthy as a marine chronometer. He conceived his solution to accurate timekeeping at sea to lie in a spring-driven clock, using a friction device to replace the seventeenth-century clock's notoriously unreliable verge escapements and the new, unproven balance spring.[73]

Like Hauton, Villons tried to deal with problems of convenience, reliability, and size. By the time he took his invention to Paris in late 1670, he had achieved notable mechanical precision: the device was no larger than a contemporary pocket watch, it had a sealed case, and it was mounted in gimbals for shipboard use. All in all, it was a remarkable device and Villons's technological achievement was significant. Unfortunately for him (and for French mariners), the device did not keep accurate time over long periods. Despite all his efforts at mechanical innovation, he had failed to produce a clock capable of solving the longitude problem.

With the aid of whiggish hindsight, Villons's problem becomes obvious. The friction device he substituted for the usual mechanical escapement depended on a paddle wheel spinning in a liquid. Such a device cannot provide long-term accuracy in timekeeping, no matter how great its mechanical precision. Without knowledge of modern fluid dynamics, however, Villons could not know that. In evaluating his work, we will find it more useful to drop concerns over ultimate technological feasibility and explore Villons's reasons for choosing this design.

Given the machining and metallurgical capabilities available to

[72]BL, A 1866, 571, 592 (13 January 1668 and 7 September 1669).

[73]See David S. Landes, *Revolution in Time: Clocks and the Making of the Modern World* (Cambridge, Mass., 1983), esp. pp. 103–144.

seventeenth-century artisans, a clock's mechanical escapement defined both its limit of accuracy and its limit of reliability.[74] Villons chose his spinning paddle wheel to avoid two clear-cut technical limits inherent in even the best available mechanical escapements. He failed to solve the problem of long-term accuracy, but he surpassed every hope for improved mechanical reliability. His device was extremely simple and employed few moving parts. Therefore, instead of suffering mechanical failure every few weeks, his clock could run for months without interruption. Even if this device was not acceptable as a marine chronometer, the Parisians at the Académie Royale recognized Villons's work as a significant technological achievement. Colbert rewarded him accordingly, with a royal appointment carrying an annual salary of more than 7,000 livres.[75] The project was a success.

It was with the fourth and last subprogram promised to Chamillart—the anatomy program—that the academy had its greatest successes. As described in the resolutions of 15 December, comparative anatomy formed part of a subprogram for biology, natural history, and medicine. Actually, when compared to what the academy promised in this subprogram, success was limited even here. "Anatomies of all sorts of animals" was only one specific resolution among several much more ambitious-sounding projects that never produced anything tangible (the herb garden and new chemical medicines, for example).[76]

The anatomy program was the mainstay of the academy's organized research, and despite continuing organizational problems, the academy proved itself extremely productive in comparative anatomy: it prepared two collected volumes of illustrated dissection reports for submission to the Académie Royale, one in 1668 and one in 1670. Moreover, the Académie Royale praised their Caennais brethren's work.

The Académie de Physique's membership possessed genuine scientific and technological talent, but Chamillart's ambitious program had little chance of overall success. Besides organizational problems stemming from his unwillingness to take on a direct leadership role, his resolutions understated the academy's real strengths while promising projects that were virtually impossible for the academicians to undertake. Filled with grand-sounding phrases, Chamillart's pro-

[74]Ibid., pp. 134–144.
[75]BL, A 1866, 655 [late 1670].
[76]Ibid., 570 (30 December 1667).

gram was not grounded in the reality of the academy's capabilities. Even if the academicians did nominally possess the technical expertise to carry out a good number of his projects (as they obviously did), the document created unrealistic expectations. Surely, anyone who really understood the academy recognized as much.

HUET'S RETURN AS *CHEF*

News of the program that Chamillart had led the academicians to formulate on 15 December disturbed Pierre-Daniel Huet. Indeed, we owe the existence of a complete list of the resolutions to him. Initially Graindorge furnished his patron with only the glowing report of "good intentions and grand projects." His letter gave very little detail on the new proposals, and Huet wanted to know exactly what the academicians had pledged. Sheepishly Graindorge complied by sending "le général et le particulier des resolutions." He had held back information, he said, only "out of fear of annoying you."[77] Clearly, then, Graindorge also recognized that there was something problematic about the resolutions Chamillart had extracted from the academicians. The intendant's fine speech must have led him to lose his senses for the moment.

The academy's next session (22 December), however, brought Graindorge back to reality. Once again Chamillart failed to appear, and once again Graindorge began to worry: "The intendant did not come to the academy on Thursday, and we will not rest easy if he comes too often because all those beautiful projects are much easier to conceive than to execute. And since they cannot be done without funds, no one feels pressure to put a hand to the work. Since the intendant was not there, M. Cally talked about the phenomenon of motion.[78] This account makes two points very clearly: first, despite inspiring speeches, nothing on the intendant's program would be done until he furnished direction (and some new funding); second, Pierre Cally spoke "on motion" explicitly because the intendant did not attend. The implied demand for Chamillart's leadership had already become a consistent theme for the academicians, but the second point presents something entirely new. Chamillart was losing what little hold he had on the academy's loyalty.

[77]Ibid., 569, 570 (19 and 30 December 1667).
[78]Ibid., 704 (26 December 1667).

Previously Graindorge had described sessions as "abandoned" or "vague" when the intendant failed to appear. Nevertheless, the academicians in attendance had always spent time discussing projects (such as the globes) that they thought Chamillart wanted. From this point on, however, if the intendant did not come to a session, the academicians followed their own interests.

Graindorge adopted this strategy consciously and deliberately. On 12 January 1668, for example, he reported that when the intendant arrived late at a session, the academicians had already launched themselves into their alternative program. Apparently that incident caused minor embarrassment, but it did nothing to deter Graindorge from telling Huet a few weeks later that he had prepared another such session: he would offer the academicians a new mathematical demonstration if the intendant failed to appear.[79]

With this system, the academy obviously could not develop the coordinated, sustained program Graindorge envisioned, yet he forged ahead anyway. We really need not look too deeply in order to understand why. Cally's discussion of "motion" was not the kind of activity Graindorge thought the academy should pursue; he wanted *expériences.* Unfortunately, he had previously gotten *expériences* only when Huet was present. Allowing the academicians to chase off in various directions with vague discourses on natural philosophy was not his idea of how to run an academy. In effect, it was a return to the format followed in the early *assemblée.* It did not put the academicians to work in the laboratory, but it did bring at least part of the membership together at Huet's house. If so, why not allow them to have their way? There was really nothing else he could do.

The session on 15 December had a paradoxical effect on the academy. After the first burst of enthusiasm died down, Graindorge became more pessimistic about the future than ever before. Indeed, with Chamillart's failure to attend the next week, Graindorge's letter took on a cynical and bitter tone toward the intendant, a tone his subsequent letter never lost.[80] From this point on he seems to have given up hope that Chamillart might act as *chef* in any meaningful sense.

Once again Chamillart had acted as a royal administrator. Graindorge even told Huet what the intendant was doing as he extracted resolutions on a new program from the academicians: he was gather-

[79]Ibid., 571, 692 (12 January and 9 February 1668).
[80]Ibid., 704 (26 December 1667).

ing information to send to "the court." In this case, Graindorge failed to understand that the "court" meant the minister Colbert, and that the resolutions were the projects Chamillart had asked for in November. Chamillart had constructed another document for administrative purposes. The resolutions made at the session on 15 December completed the paperwork he needed for his royal incorporation.

As with the statutes, we find the crux to this matter when we ask: Who pledged themselves to fulfill those resolutions? Obviously Graindorge and the academicians thought their new "patron" Chamillart had finally taken responsibility for the academy. Just as obviously Chamillart thought the academicians had finally fulfilled the charge he gave them six weeks earlier. As with the statutes, then, Chamillart and the academy had entered into a "contract" without spelling out who had responsibility for executing its terms. Intentionally or unintentionally, Chamillart had misled the academicians once again.

Chamillart deserves some credit, however. He may have negotiated in a sloppy fashion, but as a royal bureaucrat he acted with remarkable dispatch in handling the academy's administrative incorporation. Overall, in fact, his behavior belies the monarchy's reputation for slow, deliberate action. On the day he put the academy's statutes in final form, he charged the academicians with formulating a new program. When they still had not completed that task after six weeks, he went to just one session and came away with the document he wanted drafted. Then he sent off his paperwork to Colbert.

Like Chamillart, the minister Colbert also acted with an alacrity that students of the ancien régime should note. In less than three weeks he produced a response for Chamillart and the new royal academicians in Caen. He had seen the king, he had put the academy's case before the Académie Royale des Sciences, and he had decided on the formula for establishing the academy's funding. On 4 January 1668, Graindorge informed Huet, Chamillart gave Colbert's news to the academy:

> The intendant, who arrived here Tuesday night, assembled us at his house on Wednesday morning to show us a letter from M. Colbert. M. Colbert says that the king approves our academy and that it is appropriate that we should communicate with the academy that the king has established to meet in his library. M. Carcavi has written the same thing, so that by this reciprocal communication [both academies] can be made more useful to the public. After we applauded this approbation from the king, we discussed the subject of weights and measures. . . .

> We live in a time when it is not enough to have beautiful ideas, we want to put them into practice. Everything useful and glorious will no doubt be done.[81]

Louis XIV, Colbert, and the Académie Royale des Sciences had publicly affirmed Chamillart's program and his reorganization, but if the intendant truly intended to change the academy, he had produced only part of what he needed. Despite Graindorge's brave words about "everything useful and glorious," the academy was in trouble.

In many respects Colbert and Chamillart had made the task of reorganizing the Académie de Physique more difficult. The royal announcement had formalized the existence of a royal institution while placing it under obligation to open communications with the Académie Royale. This did nothing to address the academy's basic operational problem. Moreover, Colbert proved unwilling to provide funding immediately.[82] Indeed, he made all hope for funding contingent on a demonstration of the academy's potential. The only real change was that a local problem suddenly became a matter of much larger concern. With Louis XIV's *approbation,* the academy had become part of the king's *gloire.*

For the academicians in Caen, this situation was extremely uncomfortable. Chamillart urged Graindorge to force them to produce something—almost anything—to send to Paris. In other words, Chamillart once again demanded that Graindorge act as *chef.* Under the circumstances, Graindorge had less chance of filling that role than ever before. But the king himself had recognized the academy, and it had become imperative to present something of substance in Paris. Until that was done, no more favors (funding) would be forthcoming. The academy was at an impasse. Graindorge summed up the situation succinctly: "M. l'Intendant wants us to produce something worthwhile in order to attract favor, and we want to receive some dew in order to produce fruit."[83]

Between November 1667 and January 1668, then, royal incorporation had fulfilled no one's expectations. From the perspective of Graindorge and the other academicians in Caen, the situation seemed worse than it had ever been under Huet. Not only had Chamillart

[81]Ibid., 703 (6 January 1668).

[82]Although Graindorge did not specifically address this issue in reporting the royal incorporation to Huet (presumably Huet already knew), his letters through the remainder of January 1668 make the point amply clear (ibid., 571, 572, 573).

[83]Ibid., 573 (27 January 1668).

failed to replace Huet but they now had to provide some tangible response to royal "favors." In effect, Chamillart had led them into a crisis. Royal policy offered no way out.

At this point Huet finally decided to act on his concerns about Chamillart's reorganization. He did not want the academy committed to the intendant's program. What he decided was extraordinary, and he wasted no time in acting on his decision. He returned the academy immediately to the program he and Graindorge had established: they would do empirical scientific research, particularly dissections. The way he reinstated his program shows just how far Chamillart had gone in changing the Académie de Physique. The future of this royal academy would be decided in Paris, not in Caen. As Huet had moved from Rouen to Paris during December, he was strategically placed to act.

Colbert expected a formal response to Louis XIV's *approbation* from the academy. Furthermore, that response had to include a priority list for the projects in the academy's program. Indeed, under the conditions Colbert laid down, he would not decide on the question of funding until he had a better sense of what the academy could do.[84] This was the time when the academy had to commit itself to specific projects. Moreover, since satisfactory completion of some token projects had become the basis for the decision on funding, whoever authored that response to Colbert controlled the academy's future.

Huet knew that both Graindorge and Chamillart were drafting statements for Colbert in the first weeks of January.[85] Nevertheless, he was in Paris, he was still financing the academy's operation, and he simply went to Colbert and presented his own response to the king's *approbation.* In effect, he reappointed himself the academy's spokesman. Then, armed with an introduction from Colbert, he went to the Académie Royale des Sciences and outlined the program that he asserted the Académie de Physique was going to follow in the coming months: it would conduct a series of dissections on marine animals and pursue Hauton's project for sweetening seawater.[86] Of all Chamillart's grand projects, Huet was willing to support these two.

Once Huet made the academy's commitment, he wrote to Caen telling them what he had promised. In the last week of January Graindorge told him how the intendant had received this news of a

[84]Ibid., 571, 572 (13 and 16 January 1668).
[85]Ibid., 572 (16 January 1668).
[86]AdS, Reg, vol. 1, ff. 248–249.

fait accompli: "Yesterday I read portions of your letter to the academy (the parts that were proper for public reading); I read it again when the intendant arrived. . . . On his order, I also read out loud your letter to him. He approved your conduct and consented to the suppression of the letters from us."[87] Huet gutted the intendant's program. Moreover, he struck a serious blow to Chamillart's standing within the academy. Less than three months after turning the group over to the intendant, Huet had reinstalled himself as *chef*.

Graindorge probably knew what Huet was planning to do in Paris.[88] He may even have connived in the process by delaying Chamillart's efforts to draft a response to Colbert. Whatever his role, he was certainly pleased with the outcome. His patron had become the academy's *chef* once again. Not surprisingly, his attitude toward Chamillart hardened even further during January 1668. He no longer looked to the intendant to animate sessions. On the contrary, in Graindorge's eyes the intendant had become a nuisance. During the month of January, for example, Graindorge called Chamillart "un pélerin." Exactly which specific "scavenger" (a pilgrim, a shark, a locust) Graindorge had in mind is unclear, but his meaning is unmistakable: Chamillart had become a plague on the academy. By the time Graindorge reported the intendant's response to Huet's actions in Paris, Chamillart was a pest "tormenting everyone to put a hand to the work, which will hasten nothing."[89]

In effect, just as Chamillart had incorporated the academy through administrative action, Huet used the same mechanism to carry out his coup during January 1668. Acting through the intendant's own channels in Paris, he had deposed the academy's *chef.* Surely this was a questionable action: as soon as Chamillart had arranged the king's *approbation,* Huet deposed him. Depending on whose point of view is adopted, the events can be interpreted in various ways. From Chamillart's perspective, one might say that Huet simply used the intendant to gain royal favors, then abandoned his plans. In that case, Huet was guilty of crass opportunism in allowing the intendant to arrange Louis XIV's *approbation* under false pretenses. From Huet's

[87]BL, A 1866, 573 (27 January 1668).

[88]As early as the last week of December 1667, Graindorge knew that Huet was displeased with what Chamillart had done with the academy. On 30 December 1667 he wrote to Huet: "Enfin tout ce qui presentera ou il s'agira de vos interests je tacheray d'y agir avec zele et avec prudence" (ibid., 570).

[89]Ibid., 573 (27 January 1668).

perspective, however, the intendant was guilty of arranging the incorporation under questionable terms. He had manufactured the program he sent Colbert by whipping up false enthusiasm for overly ambitious and impossibly expensive public-works projects. Under the conditions Colbert laid down, the only reasonable hope for the academy's future lay in a return to the things it knew how to do best—and could afford. In that case, Huet was the pragmatist who stepped in at the crucial moment to restore a sense of reality to plans for the academy's future.

Both of those interpretations contain elements of truth, but neither captures the essence of what happened between November 1667 and February 1668. Something more than the scientific program was at stake. The real issue focused on the group's organizational form—traditional patronage versus a bureaucratically administered institution. All of the academy's problems stemmed from the fact that it had no organizational momentum when Huet turned it over to Chamillart. Graindorge and Huet had given the academy a research potential by November 1667; Huet agreed to continue his financial support; and Chamillart easily obtained corporate legitimation. Nevertheless, the academy still faced a crisis. The crux of that crisis lay in the "operational" function of patronage. The reason Huet and Graindorge wanted Chamillart's involvement was that the organization was not a viable entity. Despite their successes during 1667, the academy was not firmly established as long as its collective rationale focused on the person of the patron/*chef*. If a patronage academy was to be transformed into a scientific academy, its organizational mission had to shift to the research program.

For Chamillart's part, as paradoxical as his surrender to Huet may seem, there was a consistency to his actions. His behavior, in fact, was typical of the way royal officials created royal academies in the seventeenth century. He was acting to co-opt the Académie de Physique into state service. Certainly Chamillart wanted changes in the academy's program, but even more important, he wanted to bring the academy to life as a royal corporation. For him the legal fiction of the corporate personality was what counted—from the first discussion with Graindorge over the use of the term *royale* through his acquiescence to Huet's insubordination.

In a very real sense, then, Huet's usurpation of the academy's program was a culmination of the issues involved in Graindorge's early disagreement with Chamillart over the use of the term *royale*.

Those issues focused on the difference between private patronage and royal incorporation. There were two distinct perceptions of the process the academy was undergoing: the academicians saw a change in patrons; Chamillart and the monarchy saw the process as the co-optation of a functioning research institute. In the end, neither side achieved its goal. The net effect of the royal incorporation was to make the academy more dependent on Huet than ever before.

5

The Royal Academy of Sciences in Caen, 1668–1669

Huet's interview with Colbert produced a new situation for the royal academy in Caen. Huet promised marine dissections and the desalinization apparatus; in return, Colbert allowed him to reestablish control over the group. Nothing more was to be done about royal financing until the academy completed both halves of Huet's program. He was to pay for the anatomy program (just as he had always done), and he also promised to finance Hauton's seawater project.[1] Once those projects were done, Colbert would review the academy's situation and decide on appropriate funding. The net effect, then, was to give the new royal organization a probationary (or contingent) contract for funding.

In practice, Huet had bought time for the academy to respond to the "favors" already granted in the royal incorporation. That was important. Even more significant, however, the probationary period gave Huet and Graindorge another chance to wean the academicians from their insistence on weekly sessions with a *grand personnage.* As always, the problem was to generate a sustained research effort, one that the academicians would then maintain on their own initiative and under Graindorge's direction. If the joys of Graindorge's *curiosités* could not galvanize them, maybe Huet's ability to help them earn royal *gratifications* would.

Huet's conversation with Colbert inaugurated a new era in the Académie de Physique's history. Over the next two years, 1668–1669,

[1]BL, A 1866, 573 (27 January 1668).

the academicians fulfilled Huet's pledges for their program, producing both the promised natural history of marine animals and Hauton's desalinization project. Then, at the end of that period Colbert met his half of the bargain both by granting the academy 2,500 livres in funding for the year 1670 and by promising annual grants thereafter. Moreover, at various points during this probationary period, the academicians showed remarkable dedication to their academy. In mid-1668, for instance, Graindorge served effectively for a time as the organization's *chef,* even leading the group in the definition of new projects. Yet this was also an extremely troubled period in the academy's history. Graindorge's ability to galvanize the organization proved temporary, and twice Graindorge came to believe that the entire endeavor was going to have to be abandoned.

Overall, the years 1668 and 1669 bracketed the most intriguing and complex period in the Académie de Physique's history. On the one hand, the productivity and enthusiasm shown during the good times argues strongly that visions of a royal research institute in provincial Caen at least had a basis in the academy's working potential. On the other hand, the trials and bickerings when things went poorly demonstrate that the academy still faced substantial obstacles to establishing itself as a truly productive royal organization.

By the end of 1669 the Académie de Physique had completed its probationary period and was in line for royal funding. In that sense, the probationary period was a success. Beneath that accomplishment, however, there were signs that this trial period marked an escalation of the academy's difficulties.

SCIENCE IN THE ROYAL ACADEMY

Pierre-Daniel Huet returned to Caen in mid-March 1668. Although Graindorge wrote only two letters between the end of January and Huet's return, both show clearly that the academy had changed dramatically from just a month earlier. In the first, Graindorge summarized some of the academy's recent dissections and then described the problems created when the intendant came to sessions. His comments in the second letter succinctly indicate that the academy once again belonged to Huet: "You could not give us any more agreeable news than that of your return. I assure you that we impatiently await you to animate our academy, which has been slow these last two times although the intendant was there [last week]. For the most part, everyone is dispersed." No longer was Chamillart's presence at a session enough to animate the group. According to Graindorge's letters, the intendant had suffered a serious loss of face and his influence over the group had all but disappeared. Graindorge was not even making a pretense of trying to implement Chamillart's program of public works. Instead, he had already begun the marine dissections that Huet had promised in Paris: "We have had the good fortune to obtain a rare fish that is not given a lifelike presentation in any of our sources."[2]

When Huet did return, the academy started to work in earnest. Of course, "obtaining a rare fish" was not always possible. Henri Justel's letters to Henry Oldenburg reveal that in the first weeks after Huet's return to Caen, the academy dissected such diverse creatures as a porcupine, a mole, and a sparrow hawk.[3] The important fact is not what was being dissected, however; what mattered was that the academicians were busy at the dissecting table. Huet's presence was giving continuity and purpose to the weekly sessions. As a result, news of the activities at this newest royal academy spread quickly.

Henry Oldenburg considered the opening of the royal academy in Caen an important event, and Justel kept him informed about what the Académie de Physique was doing during this critical period. When Justel told him, "Monsieur Colbert has promised funds to the Academy in Caen for its experiments," Oldenburg began to worry that such state support would soon allow these new French institutions to outstrip English science. He warned Robert Boyle: "This, methinks, should be lodged and improved in a proper place here; but that we want zeal and industry, and consequently must needs fall under them in a short time."[4]

From what Justel was telling him, Oldenburg thought the English had good reason to worry about the new French institutions. According to Justel, both royal academies were actively engaged in new work. Oldenburg was thus able to tell Boyle: "The Parisians have lately made a dissection of a camel, where they say they have observed no inconsiderable particulars. And those of Caen have made the dissection of the eye of a hawk, and of the body of an oyster, which, I find, are likely to be published before long." Justel's next letter gave

[2]BL, A 1866, 692, 693 (9 and 17 February 1668).

[3]Justel to Oldenburg, 11 and 18 March 1667–1668, *CHO,* 4:244–245, 255–257.

[4]Oldenburg to Boyle, 30 March 1668, *CHO,* 4:282.

Oldenburg even more reason to be impressed by the science of Louis XIV's France: "The academy in Caen has had a grant from the king to build a laboratory and an observatory. They hope to be talked about soon. Monsieur Colbert has laid the first stone for the observatory in Paris, which will be magnificent. They are going to work in all earnestness."[5]

Although Justel was mistaken in believing that the Académie de Physique had been given royal funding, the essential point he was making was correct. The academy was busily engaged in its program of dissections, and it was "making noise" and "being talked about." In June the academy reported to the Académie Royale des Sciences that it may be able to supply a solution to the longitude problem. Then, in July, Huet began to fulfill his promise of reports on marine dissections by sending a long account of a particularly important one to the Académie Royale. Comments made about it in the *procès verbaux* of the Académie Royale demonstrate the eagerness with which Caen's marine dissections were received in Paris:

> A report sent by M. Huet on the dissection of a sturgeon, made by the gentlemen of the academy in Caen on the 29th of this June, was read to the company. This account reported that the sturgeon was almost six feet long. It was already a bit rotted, and its eyes had been eaten by other fish, [but M. Huet reported]: "The foul odor did not prevent us from making an accurate dissection."

The Parisians rarely had the opportunity to hear about the dissection of marine animals, and they had no access to fresh specimens for their own dissections. Therefore, Huet's complete account of the dissection of this partially rotted sturgeon was read aloud to the gathered academy in Paris. Then "the company . . . decided that M. Gallois, secretary of the academy, should write to the gentlemen of Caen to thank them for the account they sent. And since their position near the sea easily furnishes them with all kinds of fishes, [he should] implore them to continue their dissections." In response to Gallois's letter, Chamillart, as the academy's official spokesman, dutifully notified the Académie Royale that the academy in Caen would indeed continue its program of marine dissections.[6]

Huet's return to the position of patron/*chef* thus had a dramatic

[5]Ibid.; Justel to Oldenburg, 15 April 1668, *CHO,* 4:321.
[6]AdS, Reg (Mathématiques), vol. 3, f. 53v; (Physique), vol. 4, ff. 113v–114r, 116v, 166v.

influence on both the productivity and the public image of the academy. He brought the group through the critical period immediately following Colbert's announcement of royal incorporation. In mid-1668 the academy and its work were being talked about in both Paris and London. By August the academy had made significant progress toward the completion of a new treatise on the natural history of marine animals. Despite the awkward situation created early in the year by the terms of the royal incorporation, Huet's influence had worked wonders.

On 13 August 1668, however, Huet left Caen for Paris. He would not return to Caen for almost two years (15 May 1670).[7] Huet's long absence was going to put Graindorge in a difficult position once again. Huet had given the research program momentum, but he still remained the real focus for the academy's activity. Despite the appearances of what Graindorge was able to do with the academy for a time, royal incorporation had done nothing to alter that basic fact. Indeed, after Huet had removed Chamillart's influence, his success in "animating" the academy had only intensified the importance of his role.

THE DYNAMICS OF FAILURE

At the time Huet left for Paris, the Académie de Physique was enjoying the most productive period in its history. In addition to its ten statutory academicians, the anatomist Busnel, and two supernumeraries, the academy had integrated two new participants into its activities. The first was a chemical apothecary named Daleau, the other an artist named M. Du Fresne. According to Graindorge's description of these men, neither was a full-fledged academician. Daleau had requested admittance to Huet's group in early 1667, but Graindorge, who did not trust the man, consistently advised his patron against allowing Daleau to participate.[8] During 1668 Huet selected him to fill the role of *ouvrier-machiniste* called for in the statutes. His task was to build and operate a chemical furnace.[9] Almost certainly the artist Du Fresne was paid for his work of illustrating the group's dissections, and Graindorge looked forward to the time when he could produce

[7]BN, Fr, n.a., 1197.
[8]BL, A 1866, 574, 680 (20 August 1668 and 14 May 1667).
[9]See chap. 6 below.

scaled drawings to illustrate the academy's natural history of marine animals. He even planned to repeat earlier dissections so Du Fresne could furnish a complete set. With the addition of these two, then, the academy nominally claimed the services of sixteen men—fifteen of whom were working on its program until Huet left.[10]

Things continued well enough for a time after Huet's departure. Only the academicians Lasson and Villons and the technician Daleau immediately stopped attending sessions. The others kept busy. Indeed, Graindorge's letters after Huet's departure indicate that the academy was highly "animated" in August and September 1668. Chasles had produced what both he and Graindorge considered a final draft of his treatise on weights and measures; Hauton was busy with the desalinization apparatus; and the anatomists were trying to complete their illustrated natural history of marine animals. In addition, the surgeon Busnel was regularly doing special anatomy demonstrations for the group. During the third week of September, for example, he managed a rare treat when he obtained a human fetus for dissection. Even Savary was working on his translations from the *Philosophical Transactions.*[11]

Thus Graindorge was in an optimistic mood in the period immediately following Huet's departure. Everything was going well. The academy's success had even made Graindorge something of a local celebrity, a man who dealt with *grands personnages* almost on a daily basis. Following on the success of the dissection report sent to Paris, he had opened what promised to become a regular correspondence with the abbé Gallois, the acting secretary of the Académie Royale des Sciences. Graindorge was thrilled by this prospect, especially after Gallois told him that Colbert himself had read the first of his letters. Moreover, Jean-Baptiste Du Hamel, the permanent secretary of the Académie Royale des Sciences, took time to visit Graindorge and the academy as he passed through Normandy on his way to England.[12] Finally, during August and September, Graindorge was negotiating with the lieutenant-general of Lower Normandy for a subvention to help with the costs of publishing the academy's new work on marine animals.

Mixing in those kinds of circles, Graindorge had temporarily acquired the *état* he needed to bring the academicians into the labora-

[10]Chamillart, ten academicians (including Huet), two technicians, two supernumeraries, and the artist Du Fresne.

[11]BL, A 1866, 576, 577 (24 September and 26 October 1668).

[12]Ibid., 574, 575 (20 August and 6 September 1668).

tory. He expected to be able to give the intendant Chamillart some of the academy's finished work in the very near future. The academicians were even talking of some new work. Graindorge expected the academicians Cally and Vavasseur to begin a mapping project, La Ducquerie and Postel were once again at work planning the herb garden, and Graindorge thought some of the members might try a hand at building microscopes.[13]

Of course, there was some minor grumbling among the academicians. Graindorge thought some sessions were poorly attended, and it bothered him when Vaucouleurs claimed it was "time for a vacation." Such complaints must be interpreted carefully. Graindorge's standards were now so high that he considered any session poorly attended if fewer than six were there. As Huet was in Paris and Villons and Lasson were not coming to the academy, that meant he was not really satisfied unless he had regular attendance from all the other academicians. Unfortunately, like Huet, some of them also had obligations that occasionally took them away from home. Comparison with other periods of Huet's absences offers a much better basis for judging the academicians' attendance in the late summer of 1668. Graindorge never found himself alone at Huet's house, and only once did he report an attendance as small as four.[14]

Judged by the academicians' earlier record, the academy was now flourishing under Graindorge's direction. Exactly how he had gained his new credibility is unclear, but the significance of what happened one Thursday in August when Graindorge left a session early is unmistakable. After his departure, someone had brought some rare anatomy specimens to Huet's house:

> Our gentlemen of the academy came to find me, bringing M. Du C. . . . with them. He had come from Trouard with two dead snakes, which he had carried in his hand by tying them up with a ribbon. The one was a male and the other was a pregnant female, which we opened in haste without the surgeon or any instruments other than a pair of scissors. We found ten snakes in the belly of the mother. Each one was 7 inches long.

Graindorge then proceeded to give Huet a detailed account of the dissections.[15] The scientific result of the episode is less important

[13]Ibid., 691 [early September 1668].

[14]Ibid.

[15]Ibid., 574 (20 August 1668). The *curiosités* he found with this dissection then led to a new series of exchanges with Huet on Redi's work.

than what it tells us about Graindorge's standing with the academicians during this time. Not only were they attending sessions but they were excited enough about the prospect of cutting up these snakes to seek out Graindorge at his house. As a group, their attitude toward him was far different from what it had been the previous November, when half of them had walked out when he started his dissection. At this point he possessed the *état* necessary to direct the academy.

This was the kind of enthusiasm that success demanded. The academicians were even working outside the regularly scheduled Thursday meetings. Graindorge could expect to find some of them busy at Huet's house almost any day of the week. If they were not there, they might be dissecting at the tower of Chatimoine, where they were working in order to spare Huet's neighbors the noxious smells that accompanied their dissections during the late-summer heat. Graindorge's letter of 6 September 1668 describes activity during this period in a way that suggests that the academy was genuinely beginning to function in the ways he had envisioned:

> I had asked M. de Vaucouleurs for some reports we had written on a dissection of a monk fish, which he had not yet edited. [When I went to the academy last Monday afternoon, I was told they were done.] I left right away. I thought it would be good to have our artist [supply us with illustrations of the fish]. I went to the house of M. de Vaucouleurs, where I was told he was at the tower. I went there and found our gentlemen with [the artist] Du Fresne, who was [already busy] drawing the fish.

With such initiative behind the project, Graindorge had the illustrated treatise on the natural history of marine animals prepared by the third week of September.[16]

Huet had left the academy in good condition. Just six weeks after his departure they had completed the treatise on their marine dissections. Chasles had produced a draft for his treatise on weights and measures. Hauton was still busy with his seawater project.[17] Graindorge was in direct correspondence with the Académie Royale des

[16]Ibid., 575, 576 (6 and 24 September 1668).

[17]Ibid., 574 (20 August 1668). Hauton had not yet perfected his apparatus, and although Graindorge knew other *curieux* such as the one Chamillart had brought to the academy were working on the problem, he still thought Hauton had an excellent chance of becoming the first to sweeten seawater effectively. Indeed, since the Académie Royale had recently rejected a project for sweetening seawater, Graindorge thought Hauton's chances of producing something worthwhile were excellent.

Sciences, and Colbert had been reading his reports to Gallois. The academicians were even coming to sessions and planning new projects. In short, the academy finally appeared to be functioning as a scientific organization rather than as a patronage circle.

Within a short time, however, the optimism of September turned to despair. Unfortunately, Graindorge's letters offer no direct explanation for the change, but they do provide strong circumstantial evidence suggesting that Paris had rejected both Chasles's treatise on weights and measures and the natural history of marine animals.[18] Of those two possible problems, the one that would have had the most devastating effect on Graindorge (and his standing with the academicians) would have been a criticism of the work in anatomy. Indeed, since Graindorge's letters abruptly dropped all discussion of the manuscript, some problem with that work was almost certainly what brought activity to a halt in early October 1668.

Whatever the reasons for the change, sometime in early October the academicians stopped attending sessions. Graindorge himself stopped going to Huet's house, and until late October he did not even write to Huet. Henri Justel wrote to Oldenburg: "As for news of the academy in Caen, I do not know what they are doing. Monsieur Huet is here—to whom I shall give your greetings."[19] Justel's source for news of the academy was Huet, and if Huet had provided no information to send to Oldenburg, it is safe to assume that the academy was completely inactive.

At the end of October, Graindorge resumed his correspondence with Huet. He explained his failure to write by telling Huet he had suffered an eye infection so severe that it had required the application

[18]It is clear that during September 1668 Graindorge considered Chasles's treatise virtually completed. Indeed, he helped Chasles edit the final draft (ibid.). It is equally clear that within a few months Chasles was patiently at work on an entirely new version of the work (ibid., 581 [24 December 1668]). Obviously, someone other than Graindorge had told him that the original treatise was unsatisfactory. The most likely source of such a rebuff was the Académie Royale, as the treatise was known in Paris (ibid., 574 [20 August 1668]). Likewise, in the third week of September Graindorge considered the treatise on marine animals finished. After that point, he never mentioned the work again. It seems reasonable to assume that a large part of Graindorge's new pessimism was related to some difficulty with that work. Most likely, Chamillart had told them that Colbert would not furnish funds on the basis of what they had done. It is also probable that they had been told that they had to submit the work to the Paris Academy of Sciences for its opinion, and then received an unfavorable response to it. That possibility is supported by Graindorge's silence in regard to correspondence with Gallois after September.

[19]Justel to Oldenburg, *CHO*, 5:121.

of leeches three times. Probably this illness explains Graindorge's failure to write, but it cannot explain some other bad news. Graindorge had once again lost his ability to bring the academicians together at Huet's house: "If we do not take another path, we shall accomplish nothing. No one came to your house yesterday. I was the only one to go there after inviting everyone I could find."[20] The enthusiasm of August and September was gone. Graindorge had been discredited as *chef;* the academicians had returned to their old ways. That news seems to have made Huet pessimistic about the future of the academy, and he confided as much to Justel. Justel then relayed the news to Oldenburg:

> I just came from seeing M. Huet, who sends you his greetings. The academy in Caen is not succeeding. It is dispersing. In order to be a philosopher, one must be unoccupied and have no business, which is very rare in France, where life is tumultuous and full of encumbrances. In short, one must expect nothing more from this academy.[21]

Just three months after Huet's departure from Caen, then, the academy had collapsed once again.

In January 1668 the academicians had staked their hopes for royal funds on Huet's leadership. They had been productive while he was in Caen. Still, no royal money had appeared. Now, in November 1668, the academicians had become discouraged, and with their defection Huet gave up hope. At that point the academy was in serious trouble. If Huet gave up the effort, there would seem to be little chance that Colbert would supply funds.

This time under Graindorge's direction (August through October 1668) is the most curious and intriguing period in the Académie de Physique's history. Besides the tantalizing (but apparently unresolvable) questions it raises about exactly what happened with the natural history of marine animals and Chasles's treatise, this episode more than any other makes one ask why the dynamics of this organization were so heavily dependent on the personal *état* of its *chef.* On that score, this brief period under the Graindorge's direction offers crucial information.

The significance of what happened at the academy between August and late October 1668 can be judged only against events during the

[20]BL, A 1866, 577 (26 October 1668).
[21]Justel to Oldenburg, *CHO,* 5:128–129.

two earlier periods when Graindorge had acted as the organization's *chef*. During those times (most of 1666 and the first months under Chamillart) Graindorge had had difficulty bringing even one or two of the group's members to Huet's house on Thursday evenings. He had never been able to galvanize them to support a sustained research effort. Against that background, the respect most of the academicians showed for his leadership during August and September 1668 was extraordinary. Moreover, viewed against that history, the academy's collapse during October 1668 was a return to normal. Thus the episode reveals a great deal about the Graindorge's inability to become the academy's *chef*.

The October collapse tells us that Graindorge's attempt to become the academy's *chef* produced a no-win situation. The academicians demanded a *chef* who possessed a high personal *état* outside laboratory science because the activity in and of itself did not confer status in their world.[22] During 1668 Huet's arrangement with Colbert had created an artificial situation. For the first time, Graindorge's program of *curiosités*, of which the treatise on marine animals was certainly a prime example, had attracted attention from quarters that were sources of status, prestige, and *état*. Graindorge was in correspondence with Gallois, Colbert was reading his letters, and even the lieutenant-general of Lower Normandy had become interested in the academy's work. With such outside attention focused on their activities, the academies could not help but give Graindorge respect.

Unfortunately for Graindorge, such a situation could not last. His only hold on the loyalty of the academicians seemed to follow from the interest these *grands personnages* showed in the prospective treatise on the natural history of marine animals. Such a toe-hold was too narrow for safety. Moreover, it had been Huet—not Graindorge—who had interested Colbert in the work. Graindorge had only borrowed the prestige that Colbert's interest conferred. With Graindorge in charge of the project, Colbert's support evaporated at the first sign of difficulty with the treatise on marine animals. Graindorge had no way to recover the loss. Outside of his ability to fulfill Huet's promise to Colbert, his program simply did not command respect. Graindorge had nothing to fall back on. Undoubtedly that is why Huet lost confidence in the academy's ability to survive.

[22]Their objections to Graindorge were not strictly personal. For example, he counted Lasson, who refused to attend sessions he directed, among his friends.

CHAMILLART'S REVIVAL

Giving up on the academy in November 1668 turned out to be premature. No sooner did Huet decide that his efforts at transferring the academy's direction to Graindorge had been futile than Chamillart finally managed to lend a hand by obtaining something substantial from Colbert. By doing so, he once again became a *chef* who commanded respect.

We have already dealt briefly in Chapter 4 with the academy's interest in Jacques Graindorge and his solution to the longitude problem. That interest centered on a "secret" this monk supposedly possessed. For years the monk and his "solution" to the longitude problem had been a part of local lore; yet he refused to reveal his "secret" to anyone. Prying the secret of longitudes out of Jacques Graindorge was one of the items in Chamillart's program for the Académie de Physique—even though there was growing suspicion that his method was worthless.[23] Jacques Graindorge had been invited repeatedly to present his method to the academy in Caen and had refused every time. In December 1667, getting his cousin to reveal his method had even been one of the three specific projects André Graindorge promised to undertake from Chamillart's program. In January 1668 he had tried but failed.[24] In June 1668 the academy tried again to get the monk to reveal the secret by recommending his solution to the Académie Royale, without having seen it first.[25] Still the monk refused to part with his precious knowledge. In September the academicians brought him to Caen in the hope of prying the secret from him, but he again slipped back to his abbey without revealing anything.[26]

Sometime during 1668 (probably after the failed attempt to learn the method in September) Chamillart contacted Colbert directly about Jacques Graindorge and his secret of longitudes. Colbert failed to respond to Chamillart's first letter, and in November the intendant reminded him about the project: "You have never answered me about what I had the honor of sending you concerning the monk Graindorge, who wants to present you with his work on longitudes. He awaits your orders."[27]

[23]Justel had even reported this news to Oldenburg (*CHO,* 5:121 [31 October 1668]).
[24]BL, A 1866, 569, 571 (19 December 1667, 13 January 1668).
[25]AdS, Reg (Mathématiques), vol. 3, f. 33v.
[26]BL, A 1866, 576 (24 September 1668).
[27]Chamillart to Colbert, 11 November 1668, BN, Mélanges de Colbert, 149, 389r.

That reminder produced results. In the last week of November, Chamillart received Colbert's personal invitation for Jacques Graindorge to come to Paris. The intendant was pleased. Indeed, he was so excited that he wanted to deliver Colbert's invitation immediately. Despite a raging storm that made the road to Fontenay impassable for his carriage, he set out in the rain on horseback (accompanied by André Graindorge) to deliver the invitation personally.[28]

Jacques Graindorge was also pleased to have received such an obliging letter from Colbert. Still he refused to go, arguing that the journey was too expensive. Just before Christmas 1668, however, Chamillart finally broke the impasse with the monk by promising that Colbert would pay all the expenses of the trip and the presentation. Chamillart himself furnished the fare to Paris, buying two places in the coach so the monk could make the journey in complete comfort.[29] Finally, in the first week of January 1669, after more than a year of coaxing, Jacques Graindorge left Caen to present his secret of longitudes to Colbert, which meant he would present it to the Académie Royale des Sciences. A committee then examined the work.[30]

The Académie Royale rejected the project, but in the end Colbert authorized a payment of 1,200 livres to cover Jacques Graindorge's "expenses."[31] To the academicians in Caen, such willingness to pay for prying loose the "secret of longitudes" counted far more than the rejection from the Académie Royale. This was the first royal financing for any project associated with the Académie de Physique; moreover, Chamillart's dealings with the project constituted his first real involvement with one of the academy's projects. Chamillart and Colbert's involvement with the surly monk had an extraordinary effect: by sending the monk to Paris, they saved the academy from failure in late 1668. It was more than just the personal invitation, however, that produced a change in attitudes. What appears to have had the greatest effect was Chamillart's willingness to pay for the two coach fares. He had finally done something worthy of a "patron."

[28]BL, A 1866, 694 (3 December 1668). Chamillart "invited" Graindorge to accompany him. Reluctantly Graindorge complied, but he would not make the return journey with the intendant, electing instead to ride out the remainder of the storm in Fontenay.

[29]Ibid., 581, 697 (24 December 1668 and [first week of January 1669]).

[30]AdS, Reg (Mathématiques), vol. 3, ff. 261r–276v. Jacques Graindorge made his presentation on 20 February 1669. Huygens and Picard made their report on 27 February 1669.

[31]Chamillart to Colbert, 28 February and 25 March 1669, BN, Mélanges de Colbert, 149, ff. 662–663.

In early November, on the basis of information supplied by Graindorge, Huet had decided that the academy could not succeed as a royal institution. Later the same month Graindorge wanted to break off all contact between the academy and Chamillart. In fact, even after he knew that Colbert had written to Jacques Graindorge personally, André Graindorge still wanted to sever all relations with the royal government. Yet events during December caused a dramatic change in attitude. By January 1669 Graindorge was speaking of the "reawakened academy" and giving Chamillart credit for the change.[32] His letters to Huet present an exceptionally clear picture of how and why the change came about.

In describing the session of the academy that was scheduled for the last week of November 1668, Graindorge told Huet:

> The last session of the academy was even worse than the preceding ones. I had some business [that made me late]. On my way [to the academy I stopped to see M. Savary and] I was taking him there when we met M. Chasles on the same route. [On our way to the session, we met le sieur Hauton coming from the academy.] He told us that he had found himself alone there, and that after having waited, he was going home. We held the academy in the middle of the street.
>
> M. l'Intendant, who sees very well that words will not suffice to animate us, is no longer mixing in our affairs, so that we are thus in a good position to break off our commerce with him. [This royal academy] will be dead before it is born.[33]

At the time he wrote those words, Graindorge had already ridden in the storm with Chamillart to deliver Colbert's personal invitation for Jacques Graindorge to come to Paris. In fact, he described the two events (the session of the academy held "in the middle of the street" and the trip to Fontenay) in the same letter. Even knowing of Colbert's invitation when he wrote to Huet, then, Graindorge had become so soured on Chamillart and his promises that he still wanted to "break off our commerce with him."

By itself, Colbert's invitation produced little or nothing. Even Jacques Graindorge was unwilling to go until someone paid his expenses. In fact, knowledge of Colbert's invitation seems to have had no impact whatsoever on the affairs of the academy. During December Graindorge's reports sounded as bad as ever. In describing the session on 6 December, for example, he said: "Our academy has

[32]BL, A 1866, 698 [early January 1669].
[33]Ibid., 694 (3 December 1669).

nothing but a shadow. Only three of us were there last Thursday, and having produced nothing ourselves, we discussed the news."[34]

Just before Christmas, Chamillart persuaded the monk to go to Paris. The intendant "promised him the money for his journey." Not only did this offer persuade Jacques Graindorge to go but it also reawakened the Académie de Physique. The next time Graindorge gave Huet news of the academy (the first week in January), the situation had changed dramatically. Chamillart was once again "mixing in affairs" and was definitely welcome at the sessions: "[In my last letter] I was so preoccupied with the monk that I did not remember to discuss our academy with you. It was reawakened by M. l'Intendant . . . [even Villons put aside his grudge and] came. . . . Last Thursday we assembled thinking the intendant would come to the session; he did not. . . . Daleau spoke to us about his magnet."[35] Graindorge's account, which actually describes two sessions, reads as if time had been turned back a year. The academicians were coming to the sessions because they expected Chamillart to be there. They were once again discussing his projects when he did come, and when he did not, they had an alternative agenda.

For most of 1668 Huet had earned the loyalty of the academy by being present and by controlling the program. In a sense, the year 1669 would be Chamillart's, but he would never actually control activity in the way Huet had done, in part because he was not a working academician but primarily because Graindorge was the secretary. After the experience of those first few weeks under Chamillart's direction, Graindorge never again really trusted the intendant. Nevertheless, after the success at arranging Jacques Graindorge's presentation in Paris, even Graindorge admitted that Chamillart had regained some of the authority he had lost to Huet the year before.

Chamillart managed to keep the academicians busy throughout the first half of 1669. They even started some of his public-works projects, despite the fact that he had not obtained any funding for them. Not until the fall did the old problem of nonattendance reappear, but when it did, Graindorge quickly became as gloomy as ever in describing the situation. In early November 1669, for example, Graindorge made one of his most pessimistic statements: "Our academy has fallen, indeed it is in its death shroud. It will not suffice to reawaken it; it needs to be resuscitated completely." As for the activ-

[34]Ibid., 580 (9 December 1669).
[35]Ibid., 697, 698 [end of 1668, first week of January 1669].

ities of the academy, Graindorge told Huet: "M. l'Intendant wanted to see a transfusion. I found him some subjects to give him that pleasure. As for the rest, there is nothing but [planning for] projects."[36]

Despite the dismal picture Graindorge painted, the academicians were in fact working on various projects. By this time, for instance, Villons had been working on his radical new design for a marine chronometer for almost a year—even if he seldom attended sessions. Most important, Hauton finally had his seawater apparatus perfected. In early December 1669 he made his presentation at the Académie Royale. His method received the endorsement of the Paris academy, and Hauton received a *gratification* of 1,200 livres in the name of Louis XIV.[37] Morale at the Académie de Physique was so low, however, that even the news of Hauton's success in Paris aroused little enthusiasm in Caen.[38]

Although Graindorge had once again become cynical about the intendant's promises, Chamillart was at that very time arranging funding for the academy. Hauton's seawater project was received in Paris as convincing proof of the academy's potential. In fact, with the completion of that project, the academy had actually delivered both halves of Huet's promised program (if we assume that the treatise on marine animals had been sent to Paris in October 1668). Graindorge's letters to Huet make it clear that Chamillart was asking the academy to draw up a specific budget that he could submit to Colbert. Indeed, Graindorge had told Huet as much when he reported almost offhandedly that there "was nothing but projects." Graindorge simply did not believe anything would come of it.

Chamillart was promising a new and closer relationship between the Académie de Physique and the Académie Royale des Sciences. He was, in effect, promising that the academy in Caen would become a subordinate section of the Paris academy. Graindorge was indifferent. He neither protested nor made a move to alert Huet to this turn of events. Huet apparently learned of the intendant's intentions only through the gossip in Paris. When he asked Graindorge about these matters, Graindorge responded: "When I received your letter saying that you are not of the opinion we should second the Paris academy,

[36]Ibid., 593, 594 (7 and 28 November 1669).
[37]AdS, Reg (Physique), vol. 6, ff. 189r–198r.
[38]BL, A 1866, 595 (20 December 1669).

the affair was already in the hands of the intendant—with no way to remedy things."[39]

Unlike the situation in January 1668, there was no chance for Huet to subvert Chamillart's intentions. Graindorge had not even told him what was happening. It had been two years since Chamillart had become *chef,* and despite the academy's efforts in 1668 and 1669, no substantial assistance had appeared from Paris. It seems Graindorge simply did not believe the intendant would obtain anything after all the time that had passed. He saw no reason to orchestrate opposition to anything the intendant might try to do.

Graindorge was wrong. On 16 January 1670 Chamillart announced that Colbert had decided on funding for the academy. Moreover, the academy in Caen was to have a close association with the Paris Academy of Sciences. Graindorge told Huet what had happened at the session at which Chamillart announced these things:

> Yesterday M. l'Intendant did not fail to come from Bayeux straight to your house [for the session of the academy]. He was the first one there. I arrived soon after, and when the company was numerous, he told us the sum M. Colbert has destined for our academy, which greatly surpasses our hopes. [That 2,500 livres will] most vigorously engage us in work in order to give a response to the wonderful hopes that have been conceived for our academy. We resolved to write to M. Colbert to thank him for the favor he has shown us; also to Monsieur Carcavi to testify to our recognition of the pains he took to help us receive this *gratification,* as well as for the gracious offers of assistance he has given us.

Thus the Académie de Physique had finally become the organization Chamillart had wanted to create from the start. It was a state-supported research institute.

All the group's problems were not over, however. As Graindorge continued his report on this extraordinary session, it becomes clear that in many respects the group's troubles had really just begun:

> After that, M. l'Intendant read the part of your letter in which you made the offer of your house—not only the part you occupy, but also the part you rent out—and then he asked for the opinions of the company [on this matter]. He gave me the honor of asking mine first. I spoke of how advantageous this offer is, and said how obliged we will be to you since we will need to rent a house anyway, with no guarantee of

[39]Ibid., 595 (20 December 1669).

> finding one that has all the conveniences of yours—or one so well situated. M. Chasles spoke next. I had already told him of your plan, and he acted like a man of honor [by supporting you].

Even as Chamillart was announcing the news, then, the academy was changing in response to royal financing.

The meeting Graindorge described was held at Huet's house, just as the sessions had been held there since the days of Huet's informal *assemblée*. The academicians were seated in the only place they had ever met; yet with the announcement of royal funding, the academy was going to have to find a meeting place and equip its own laboratories. In fact, Huet's "advantageous offer" of his house as the new home for the academy amounted to a demand that the group pay rent for the space where it had been meeting for eight years.

The former patron was offering to become the new landlord. The fact that Graindorge had prepared Chasles to speak in favor of this plan indicates two things: first, Huet had advance word that Colbert was going to announce funding, and second, Huet's plan to rent his house to the academy was expected to displease some. It did. Daleau, the apothecary who had become the academy's *ouvrier-machiniste,* spoke next:

> Daleau . . . stammered . . . that the presence of a technician is necessary to our activities. [He wants to be paid for his part in the academy.] . . . He said in his embarrassing speech that he would want to work at home. . . . He would soon exhaust our funds. He says that he has broken 500 écus worth of glass in the last two years.[40]

According to Graindorge, then, Daleau not only opposed the plan to rent Huet's house ("he would want to work at home") but also was trying to swindle the academy.

Colbert had committed 2,500 livres to the Académie de Physique for 1670. Of that amount, 1,500 livres (500 écus) was to be the yearly grant. The other 1,000 livres was for past expenses as a royal academy and for the extraordinary expenses expected during the first year of royal financing. Daleau's "stammered" and "embarrassing" speech presents a number of issues related to Colbert's grant. Five hundred écus (Daleau's breakage) was exactly the amount the academy had been given for its day-to-day expenses (3 livres = 1 écu); the two years it had taken Daleau to break glass costing that sum was precisely the

[40]Ibid., 596 (17 January 1670).

amount of time that had passed since Colbert had first promised to finance the academy.

Daleau (according to Graindorge) was shamelessly trying to cheat the academy out of an entire year's grant by claiming a reimbursement for past expenses. Furthermore, with his demand for a "hired" operator (a position for which he was the obvious candidate), Daleau was trying to lay his hands on some of the remaining 1,000 livres. Graindorge was outraged by Daleau's crass behavior, but then, he was not so innocent himself. His report on having coached Chasles to speak in favor of renting Huet's house smacks of scheming to put some money into his patron's pocket.

No sooner had Chamillart finished reading Colbert's announcement than the squabbles over the money began. Huet's demand for rent and Daleau's demand for a salary and reimbursement for broken glassware became continuing sources of conflict, but the new problems were not limited to these squabbles. At this point the divided loyalties produced by Huet's patronage started to become truly significant in determining the academy's future. Now that the academy had funds, the choice between Chamillart and Huet took on new meaning. The royal funding finally gave Chamillart a legitimate claim on the academy's program. With money to commit, the real organizational decisions had to be made.

6

The Royal Academy of Sciences in Caen, 1670–1672

In his *Origines de la ville de Caen,* Pierre-Daniel Huet blamed royal funding for destroying the Académie de Physique, and he explicitly cited the academicians' greed as causing all the academy's problems. In effect, he accused them of acting childishly when they saw prospects for sudden riches.[1] Huet had good reason to see royal financing as marking the beginning of the end for the Académie de Physique, but at best, the academicians' greed is a feeble explanation for the academy's failure. If royal funding contributed to the academy's slide downward, it was not because a few individuals suddenly sought to enrich themselves. As Graindorge himself told Huet, this was "only a small problem."[2]

Even before the money arrived, the academicians argued over how to spend it. Several particularly resented Huet's demand that they pay rent for his house. Moreover, most opposed the specific purchases Graindorge wanted to make. Matters would only become worse in the fall of 1670, when Huet "resigned" the academy to Graindorge and moved to Paris. Throughout 1671 and 1672 Graindorge continued his efforts to resolve the difficulties associated with royal financing and administration, but with little success. Finally, in late 1672, Graindorge tried one last time to reach an accommodation with the royal bureaucracy when he attempted to present the academy's plight directly to Colbert. He failed. As a result, the Académie de Physique

[1]Huet, *Origines,* p. 173.

[2]BL, A 1866, 598 (10 February 1670).

closed, almost exactly five years after Chamillart had taken it under his protection.

The problem faced in efforts to adjust to royal financing was the same one that had plagued the group since 1666. Without active participation and operational direction from the patron/*chef* (whether Huet or Chamillart), the academy did not function. As soon as Chamillart and Huet left Graindorge to administer the royal funding (which, of course, they did) the academy began to fall apart. The arrival of money, then, marked the beginning of the end, but it did not cause failure. If royal financing hastened the academy's final collapse, it was only because it helped bring Huet's "patronage" to an end.

PATRONAGE UNDER ATTACK

Even as Chamillart read Colbert's letter announcing the 2,500 livres in royal funding, Huet's position in the Académie de Physique changed. The royal money ended his financial patronage. Indeed, Huet's demand for rent underscored just how dramatically royal funding altered his relation to the group. Those academicians who opposed his "advantageous offer" rejected their former patron. They thought the academy should move, either to another member's house (Vaucouleurs's or Graindorge's) or to the tower of Chatimoine. Graindorge turned aside every suggestion along those lines. The tower was unfurnished, his own house was "inconvenient," Vaucouleurs had "no right" to host the academy. Within weeks Graindorge was trying to reassure Huet that the "offer" had not totally destroyed his authority within the group, by telling him he still enjoyed support among the academy's "honest men."[3] Placed against the kind of respect Huet had always commanded in the past, such a statement can have given little comfort.

Graindorge knew that Huet's demand for rent would factionalize the academy. He also planned to defend his patron's position. Both points are amply demonstrated by his admission that he coached Chasles to support Huet's offer. Chamillart's endorsement (he gave

[3]BL, A 1866, 597 (27 January 1670). Graindorge did not list the academy's "honest men," but we can suggest who these were by eliminating those who in other contexts said or did something to oppose renting Huet's house. At best, Huet's demand for rent enjoyed support from only three members: Graindorge, Chasles, and Postel.

Graindorge the "honor" of speaking first) made Graindorge's task in this particular matter easy, but the victory proved costly. Huet had put his *fidèle* in a difficult position. Graindorge had to place his patron's "interests" above the academy's larger interests—as expressed by what most of its members wanted to do. He even refused to allow the academy to meet at his own house. In effect, Huet created a test of Graindorge's loyalties. Would Graindorge continue to support his patron come what may, or (if the academy's survival became an issue) would Graindorge throw in his lot with those who wanted to decline Huet's "advantageous offer"?

Though André Graindorge lacked the personal *état* necessary to function as the academy's *chef,* he suffered from no weakness of will or indecisiveness, especially as he faced matters of principle. He knew Huet's demand for rent had created dissension, but he knew also where his first loyalty lay:

> I have no doubt whatsoever but that jealousies will continue to run through our academy, but in the end, when someone grumbles [about your house] I just ask them to furnish a place that would be better suited. . . .
>
> In the end, they can say whatever they like, and we will go on with our plans. I shall always emphasize the obligation the company owes you in so many things.[4]

Thus, after years of trying to lure the academicians to sessions, Graindorge found himself turning aside the first real initiative they had ever shown for making the academy work. His loyalty to Huet outweighed his commitment to the academy.

Graindorge's reports on dissension within the academy caused Huet to have second thoughts about renting his house to the group. He wanted to back down on the issue and told Graindorge that he would ask Chamillart to reopen the question. Once again, Graindorge's loyalty to his patron's interests outweighed his desire to mollify the academicians. He urged Huet to ignore everything he (Graindorge) had said in recent weeks:

> I do not know how you understood what I told you because things seem not well enough explained. Everyone who was at our session was of one voice in accepting the advantageous offer of your house. I wrote to you that perhaps their inner feelings did not agree with their words,

[4]Ibid.

> but I had judged the matter by all the things I have heard in their speeches of the past, which has nothing in common with the present. There was really only the small problem with Daleau, which did not concern you so much as it concerns his selfish interests. . . .
> So I do not know on what basis you would write to the intendant asking him to reopen a matter that has already been settled.[5]

Despite Graindorge's rationalizations, he had helped maneuver the group into a commitment that most members found unsatisfactory. Even the verbal screen he threw around this unpleasantness for Huet's benefit is transparent: he judged the academicians' "inner feelings" by their "speeches in the past," but on renting Huet's house, he accepted the "one voice" (unanimous support) they gave while Chamillart listened. Given the academy's new circumstances, in which some wanted to attack Huet's authority, such high-handedness would come back to haunt Graindorge.

With Chamillart's backing, Graindorge had made the rent for Huet's house the academy's first expense. Of course, that commitment had not actually put any cash into Huet's hands. Like the academy itself, he had to wait for his money. Colbert's announcement only allocated funds. To "touch" the money, Chamillart had to obtain from Colbert an authorization for payment, which would enable Graindorge to draw the academy's money from local royal officers who handled the king's revenues. Depending on which royal revenues Colbert assigned to the academy, the wait could be considerable.[6] Moreover, Chamillart was unwilling to request immediate disbursement. He wanted a more specific budget first. The net effect, then, was that the academy did not even see any of its money until mid-1670. Such a delay proved crucial.

Chamillart wanted Graindorge to refine the academy's budget. Graindorge did so, and forged ahead with new financial commitments, almost exclusively to the benefit of the anatomy program. He was once again demonstrating his loyalty to Huet. And once again he created a situation that was bound to cause problems. It is clear that Chamillart had asked the academicians to work on some other projects: "M. l'Intendant favors an investigation of dyes; this is timely [because of his efforts to establish new textile industries here]. M. de

[5]Ibid., 598 (10 February 1670).

[6]For a more specific description of the complications possible in such a transaction, see Julian Dent, *Crisis in Finance: Crown, Financiers, and Society in Seventeenth-Century France* (New York: David & Charles, 1973), pp. 27–43, 82–83.

Villons says that he has orders to draw up reports concerning the quarry at Colleville, [the project for making] the river navigable, and the draining of some swamps." At the same session at which the academicians discussed those plans, Graindorge also presented a "list of books and instruments that are necessary for our work."[7] Nothing from his "list" contributed to Chamillart's tasks.

Obviously the prospect of spending royal money had created a complex situation. In backing the academy's "decision" to rent Huet's house, Chamillart had played a major role in silencing all opposition to what Huet and Graindorge wanted to do with the royal funding. Then he ordered the academy to plan its spending. Now, just weeks later, he suffered the consequences of his own action. The intendant clearly wanted to apply the royal funds to projects he had proposed two years earlier. Yet since he had confirmed Graindorge's authority over the royal money, none of the academicians raised a voice in protest or offered any suggestions for changes in Graindorge's budget; "not one of them said anything about it."[8] Chamillart had endorsed Graindorge's control over funding, and he would have reason to regret it.[9]

Unfortunately, the freedom Chamillart gave Graindorge in running the academy allowed Graindorge to concentrate his attention (and the academy's money) almost exclusively on the program of dissections. He even refused to allocate money for the astronomy program.[10] The academicians acquiesced in Graindorge's planned expenses. Graindorge then sent Huet a long list of books for purchase, and he weighted it overwhelmingly toward anatomy and natural history.[11] Such books were expensive, and Graindorge knew that

[7]BL, A 1866, 705 (28 February 1670).

[8]Ibid.

[9]He also had reason to support Graindorge. Although the intendant was asking for public-works projects again, he was most interested simply in ensuring that the academy produced something to show Colbert. In effect, royal funding produced a situation that was exactly the same as it had been in January 1668. The academy had been given royal "favors," and Chamillart wanted something to offer as an immediate response to Colbert. Just as in the earlier case, his best option was to rely on whatever Huet and Graindorge could give him. This time he knew exactly what to do. He was even willing to help them with the program in marine anatomy. To that end, Chamillart personally brought a dolphin to the academy and gave orders that it should be dissected immediately (ibid., 634 [before 1 April 1670]). We know he was trying to generate something for Colbert because Graindorge promised to send Huet "une relation exacte et curieuse de la dissection," but was unable to do so (ibid., 635 [before 1 April 1670]). Chamillart himself had taken the report to have it edited, copied, and sent to Paris. Chamillart, then, was willing for Graindorge to continue dissecting—if that was what produced results.

to purchase everything would require a great deal of money. When he sent his list to Huet, however, he thought the Académie Royale des Sciences could supply many of these books from duplicate copies in the king's library. That arrangement had been among the "gracious offers of assistance" Pierre Carcavi had made in January.[12] He told Huet to act accordingly: "Whatever you can obtain from the king's library will spare our purse. As for the remainder, manage as you think best because we must not overextend ourselves since we still have to do what is necessary here to meet the costs of chemistry and the anatomy of fishes."[13] When Graindorge wrote those words, he thought that Huet would need 500 livres for purchases in Paris. Graindorge's reference to that 500 livres constitutes one of the very few instances in which his letters specify an actual budgetary amount. In large part, we can explain his lack of budgetary specifics simply by the fact that he had little idea about what things would finally cost. This initial 500 livres, however, turned out to be just half of what he finally had to assign Huet for expenses in Paris. He expected to spend the remainder of the money in Caen.

The major item among local expenses involved remodeling the laboratory facilities in Huet's house. The plan called for equipping new laboratories for both chemistry and anatomy. On this subject, Chamillart insisted that Graindorge make progress. Chamillart thought the academy should now establish itself as a research institute. As Graindorge told Huet:

> M. l'Intendant asked which rooms in particular you can furnish our academy. I told him as well as I could according to what you indicated to me in one of your letters. But there is a woman lodging in your house. I was told that she says she will not move out for at least three months, or even six months if she is not notified before Notre Dame [May]. . . . As things stand with your house, we are not able to make use of anything but your own chamber, which is where we hold our sessions.[14]

[10]Ibid., 599 (1 April 1670). Initially Huet's continuing desire to establish a program for astronomical observation had led Graindorge to propose the purchase of new instruments (ibid., 634 [n.d.]). At the point at which he began to feel the royal funds were limited, however, Graindorge abandoned the astronomy program.

[11]Ibid., 634 [before 1 April 1670]. For a published transcription of Graindorge's list, see Brown, "L'Académie de Physique," pp. 175–181.

[12]Ibid., 596 (17 January 1670). Along with the offer of books, Graindorge reported that Carcavi promised assistance with the fabrication of new instruments for dissection, two microscopes, and the construction of astronomical instruments.

[13]Ibid. [before 1 April 1670].

[14]Ibid.

After forcing through the proposal to rent Huet's house, then, Graindorge learned that the extra space needed was unavailable.

Although Huet's tenant refused to move, the intendant insisted that remodeling had to begin. Graindorge reminded Huet: "You will have seen from my last letter that the intendant is demanding that we put our hands to work and prepare the [laboratory space]. . . . Send your instructions on this matter."[15] At that point all Graindorge's planning stalled. He wanted action from Huet. Huet had responsibility for arranging the purchases in Paris and for deciding what to do about the tenant in his house. From Graindorge's viewpoint, everything was beginning to go wrong.

In early April, Huet finally responded to Graindorge's requests for books and instruments, telling him these things were far more expensive than they had thought. He explained that Carcavi had withdrawn the offer to supply books from the king's library. As Graindorge said, "there was no angel to double the dose of money." Huet wanted him to establish priorities among the items the academy needed. When he did so, he explicitly confirmed the decision to make dissections the academy's cornerstone: "If it is necessary to pay immediately for the [dissection] instruments . . . , we should begin with that expense because if we have anything to recommend us it will be the anatomy of fish. Those gentlemen [at the Académie Royale] will always give us the [current] opinions on every other topic."[16] Huet gave Graindorge instructions on remodeling the laboratory facilities, but Graindorge could not carry them out because they involved Chamillart, who had just left for Paris. Graindorge did not describe Huet's instructions, but it appears Huet was unwilling to evict the woman who had rented most of his house. He wanted to accommodate her and at the same time give the academy the space it needed. With Chamillart out of town, Graindorge decided to await Huet's return to Caen before doing anything more.

By early April, when Graindorge stopped planning the new laboratories, the anatomy program constituted the academy's only organized activity. Later that month, this work too ended. An epidemic of respiratory disease swept through Caen, and in the midst of this "mortality" a professional dispute broke out among the academy's

[15]Ibid., 635 [before 1 April 1670]. For discussion of similar problems in establishing laboratories at the Académie Royale, see Joseph Schiller, "Les Laboratoires d'anatomie et de botanique à l'Académie des Sciences au XVIIe siècle," *RHS* 17 (1964): 97–114.

[16]Ibid., 599 (1 April 1670).

médicins. Their argument centered on the usefulness of venesection in treating the disease. After Hauton, the iatrochemist, let a patient die without at least trying a phlebotomy, the traditional Galenists (Vaucouleurs, La Ducquerie, and Postel) claimed that bleeding would have saved the man. When they brought this argument into the academy, Graindorge tried to silence them. He said such a debate exposed the entire group to ridicule. His peacemaking efforts did little except possibly polarize the sides even further.[17] All the participants in the feud stopped attending sessions.

The epidemic delivered its final blow when it took the life of the academician Savary.[18] With Huet and Chamillart both in Paris, one academician dead, some angry about the renting of Huet's house, others involved in an iatrochemical-Galenist feud, little could be done at the academy. Not surprisingly, the week after Chamillart left town, Graindorge reported bad news: "I have had no more help from our gentlemen than if I never saw them. No one has done any research. . . . You and I will make more progress in one day than we are making now in a month."[19] Clearly Graindorge once again believed that the academy's success depended on Huet's return.

The situation remained unchanged throughout April. At the end of the month Graindorge poured out his complaints about the inactivity in a long letter. According to Graindorge, the academicians were a selfish, stupid, lazy, and greedy lot: "We would have produced more if there had been no funds." They refused to pay for anatomy specimens; they were secretive and vainglorious; they promised miracles but wasted money while "only producing smoke." One by one, he ticked off their individual faults. He concluded his tirade in obvious exasperation: "I scream and I harass, but that is all I can do."[20]

In effect, despite his desire to blame royal financing for this latest collapse, the situation Graindorge described paralleled what had happened previously every time Huet or Chamillart had left the academy on its own for a few weeks. The promise of royal funding had made no difference.

After less than three months of planning, then, the academy stalled once again. Graindorge's insistence on renting Huet's house alienated the majority of the academicians. Huet's unwillingness to evict

[17]Ibid.
[18]Ibid., 635 [April 1670].
[19]Ibid., 635, 600 ([April] and 7 April 1670).
[20]Ibid., 601 (21 April 1670).

his tenant left the group without adequate facilities. Graindorge's purchases for the anatomy program already promised to become far more expensive than anyone had anticipated. And finally, the dispute among the *médicins* destroyed the anatomy program. Given the situation, the only solution that occurred to Graindorge was the one that had always worked in the past: to rely on Huet to straighten things out.

THE BREAKDOWN OF PATRONAGE

In early May, Chamillart obtained authorization for the academy to receive its cash. He gave Huet 1,000 livres for purchases in Paris and arranged for Graindorge to "touch" the other 1,500 livres in Caen.[21] Huet returned to Caen in mid-May, and from that point on the academy (the anatomists, at least) resumed activity. Huet made no better progress than Graindorge in constructing (or equipping) new laboratory facilities, however.

Although no documents describe the academy's activities during the summer of 1670, Huet's presence revitalized the anatomy program. In the fall Huet sent a report on his new investigations of the urinary tract in dogs to Henry Oldenburg.[22] Also during the fall Graindorge received news that the Académie Royale des Sciences planned to include recent dissections from Caen in a new volume on natural history.[23]

During the fall of 1670, however, the situation in Caen once again changed dramatically for the worse. In September Huet became *sous-précepteur du dauphin.*[24] Soon he moved to Paris, removing all the furnishings from his house in Caen. That left the academy with no suitable place to meet. A house without tables and chairs or instruments or library was of little use. In addition to Huet's withdrawal from the scene, the Académie de Physique lost another of its academicians at about the same time. Villons, arguably the academy's only

[21]Ibid., 602 (1 May 1670). Unfortunately, Graindorge's letters never reveal how they spent this money. As we shall see later in this chapter, he had no clear idea himself. He never received an accurate accounting from Huet for the 1,000 livres spent in Paris, and he had only a vague idea of how he spent the 1,500 livres entrusted to him.

[22]Huet to Oldenburg, 20 October 1670, *CHO,* 7:206–207.

[23]BL, A 1866, 603 (13 November 1670).

[24]For a description of the events leading Huet to this position, see Tolmer, *Huet,* pp. 400–403.

productive mechanician (even if he boycotted the academy whenever Graindorge tried to act as *chef*), had built a new type of clock that might serve as a marine chronometer.[25] On the basis of his work, he gained royal employment in Paris. Thus within six months the Académie de Physique lost three of its original ten academicians (Savary, who had died in the spring; Villons; and Huet). In a short letter written soon after Huet left Caen, Graindorge commented on the two most recent departures:

> M. l'Intendant arrived yesterday evening. He told me that Villons has been retained by the Academy of Paris and that he received the bonnet and the [illegible word] in the presence of the academicians, who all admired his work. The intendant also said that Villons has orders to build a clock for the king. . . . [The intendant also] said that he took no pleasure in saying these things, but that [Villons was glad] to leave our academy in general and me in particular.
>
> We have not been able to watch your possessions leaving here without emotion, and although you might try to persuade us that this is of no consequence, we cannot help but be disquieted by your long absence . . . [but] we do understand these things when we reflect on your personal interests, because when we consider your satisfaction and your advancement, it quiets all our murmurings.[26]

If Graindorge's "disquieting" thoughts involved premonitions about the future of the Académie de Physique, he had reason to worry. With Huet gone and his house empty, the academy was in for a more difficult time than ever before. In this same note Graindorge also revealed that the academicians had already stopped working: "Our academy has produced nothing."

Although Graindorge sent reports on the summer's dissections to the Académie Royale, he did little else during the fall of 1670. He even stopped writing to Huet. Not until mid-November did the academy again show signs of life, apparently because the Académie Royale had responded favorably to the dissection reports. This was a mixed blessing, however. When Graindorge learned that Colbert had ordered the Paris Academy to publish a large new volume on natural history—using some of the drawings from Caen—he was not pleased. The arrangement denied the Académie de Physique the honor of

[25]Villons had been working on this project at least since January 1669 (BL, A 1866, 698 [early January 1669]). In December 1670 he received a royal appointment "avec sept mille livres et plus d'appointment" for his efforts (ibid., 655 [late 1670]).

[26]Ibid., 702 [early fall 1670].

publishing its own work. In his mind, the Parisians were robbing the Académie de Physique.[27]

At the same time that Graindorge gave Huet this news, he also told him that Chamillart had again asked the academicians to establish their laboratories: "M. l'Intendant is urging us to find a house where we can establish the academy and build the furnace. He is offering [the academy] to Daleau and the other chemists, who will soon exhaust our small funds if they have their way. No one is willing to rent under his own name, and no one will give us a lease since the funds are very uncertain."[28] With Huet gone, Chamillart removed some of Graindorge's authority over the academy. If Graindorge could not furnish a chemical laboratory, the intendant wanted the others to build it. Graindorge reported nothing new for over a month, but then told Huet: "In haste, our academy has rented the Maison du Fauconnier at the far end of [the parish of] St. Estienne and established Daleau there. He is a great swindler, facing debts with a large family. He has duped the greater part of our academicians [into giving him money] without ever repaying them."[29] Chamillart wanted the chemists to start working, but he had not approved the rental of this house, which turned out to be less satisfactory than Huet's house had been. Even before they had rented the Maison du Fauconnier, some of the academicians had rented space for the academy at an inn.[30] That was a scandalous and degrading thing to do. When Graindorge found out about this bizarre turn of events, he decided to lay all his complaints before the intendant.

Graindorge finally succeeded in convincing Chamillart that the other academicians were squandering the group's meager resources and that the issue of a meeting place must be settled once and for all. Moreover, he persuaded Chamillart that the recent attempts to rent space demonstrated how unfit the others were to make any decisions affecting the academy. Chamillart called the group together at his house, and after a long session filled with personal recriminations and heated debate he ordered the group to meet at Huet's house once

[27]Ibid., 603 (13 November 1670).

[28]Ibid. As *chef*, Chamillart was the logical person to sign a lease, but the intendant had always proven himself unwilling to fund the academy personally, even though he more than anyone should have felt confident of being reimbursed from the promised royal funds. In light of his refusal, none of the academicians was willing to step forward as the lessee for the academy.

[29]Ibid., 604 (18 December 1670).

[30]Ibid., 655 [December 1670].

again. Graindorge immediately left the session and took possession of Huet's house.[31]

Chamillart had issued an ultimatum. He told all those who opposed Graindorge that they could either submit to his leadership or leave the academy. In doing so, he finally brought the wrangling to an end. Even more significant, such an action marked the definitive end of Huet's influence over the group. For the first time in his three years of association with the group, Chamillart had asserted his prerogatives as *chef.* In fact, his ruling appears to have stopped the disputes. None of the academicians left. Following his ultimatum, they actually made the first substantial progress toward building the new chemical furnace and outfitting the laboratories at Huet's house. Of course, this one session could not cure all ills. Indeed, the foreshadowing of future problems can be seen in the way that, even as Chamillart issued his ultimatum, he once again delegated the responsibility for action to Graindorge.

SCIENCE UNDER THE BUREAUCRACY, 1671

Despite his crucial role in ending the dispute over a meeting place, Chamillart still refused to act as the academy's *chef.* And the result was predictable. After several weeks of organizing laboratory equipment and gathering building materials for the new chemical furnace, the academicians' enthusiasm cooled and attendance fell off. Graindorge was virtually on his own, with Chamillart as remote as ever. The royal money was almost gone, and Graindorge had heard nothing about funding for 1671. The intendant never appeared at Huet's house. Finally, in desperation, Graindorge wrote to Chamillart asking what the academy could expect. The response was encouraging: "He answered me by saying that he will work to see that our academy receives the funds it was granted."[32]

With that promise from Chamillart, the academicians settled down to their old pattern of waiting for action from Paris—no funds, no new projects.[33] This time, however, they waited under new circumstances. The academy had received 2,500 livres during 1670, but

[31]Ibid., 656 [December 1670].

[32]Ibid., 632 (20 April 1671).

[33]Ibid., 659 [spring 1671]. In this letter Graindorge explicitly stated, "Nous suspendrons le travail jusques a un fonds asseuré."

those funds had not produced anything like the results promised. Colbert refused to send any more money. To check on progress, he ordered Pierre Carcavi to investigate.[34] Graindorge told Huet:

> I received a letter from M. Carcavi, who asks for a report on what each one of our academicians has done. I am going to have a great deal of difficulty in drawing up an account. Moreover, he has told M. l'Intendant that there is a division in our academy, that we no longer assemble, and that there is still no laboratory. He says that he is awaiting the intendant's instructions in order to talk to Monsieur Colbert, from whom the intendant has asked for 1,000 livres for this year according to what His Majesty has granted us—even though the king should give us 1,500 livres per year.
>
> Since we have received nothing this year, we are in no state to move on to anything that will involve expenses. Besides which, our chemists are so secretive about their imaginary discoveries that we would have trouble succeeding in any case.[35]

Leaderless and almost out of money, Graindorge had to justify the academy's existence.

Graindorge responded to Carcavi's letter with a summary of the academy's activities; he did not account for the expenses, however. Carcavi thought Graindorge's report was vague and difficult to understand; furthermore, he wanted to know where the money had gone. During the fall of 1671 he sent Graindorge a list ("*par articles*") of questions he wanted answered. Carcavi then told Graindorge not to expect any new money until he furnished answers.

Graindorge reported Carcavi's demands to Huet in detail. They read like an indictment of the academy's failures: Carcavi said the academy must give regular progress reports on its work as well as complete accounts of all its finished projects. He demanded that each academician have a definite task to work on. He said that the academy should have only three rooms—one for sessions, a laboratory for chemistry, and one for dissection—rather than the whole house it was renting. Finally, he charged that the academy had spent too much on books.[36] In effect, Carcavi insisted that the academy fulfill the contract it had entered into in November 1667 and demonstrate that it was the organization promised in the original statutes.

[34]For biographical information on Carcavi, see Charles Henry, "Pierre de Carcavy, intermèdiare de Fermat, de Pascal et de Huygens, bibliothècaire de Colbert et du Roi, directeur de l'Académie des Sciences," *Bulletino Boncampagni,* 1884, pp. 317–391.

[35]Ibid., 657 (13 August 1671).

[36]Ibid., 670 [fall 1671].

Since the academy no longer met, Graindorge had a difficult time compiling a response to Carcavi's *articles.* He rarely saw most of the academicians, and he was not familiar with their recent work (if they were doing any). Moreover, the laboratories were not in anything near the condition Carcavi's terms demanded. Worst of all, Graindorge could not account for the 2,500 livres he and Huet had received the year before. He had little idea how he had spent his 1,500 livres, and he had no idea how Huet used his 1,000 livres. Despite his claim that he gave Carcavi an "exact account, almost complete to the last detail," he did not know what had happened to the money.[37]

In the fall of 1671 Graindorge became extremely suspicious of royal officials. In particular, he began to think that the Académie Royale was trying to sabotage the Académie de Physique. Graindorge's growing paranoia was evident, for example, in the way he dealt with Pierre Carcavi. He told Huet that when he sent the academy's illustrations to Paris, he took precautions: "I sent the paintings that we had made of fish to M. le Chevalier de Villons for safe delivery into the hands of M. Carcavi so that he cannot do what he did with the last package, which he swore he never received."[38] From this point on, Graindorge's dislike and distrust for Carcavi would only grow. He perceived Carcavi as an obstacle standing between the Académie de Physique and its "patron" Colbert. According to Graindorge, if Colbert could learn the true state of affairs in Caen, he would certainly renew funding.

The academy's problems had finally begun to wear Graindorge down. Dealing with Parisian bureaucrats made him cynical. The lack of new funding made him bitter. In December he decided that the academy had to give up Huet's house. In January 1672 Graindorge's resignation and pessimism turned to indignation. He thought that the academy had satisfied the demands from Paris and that the funding should resume: "M. Carcavi wrote to me that he had an order from M. Colbert that we account for our work and our expenses. It is a simple matter to see that they have made various demands of us, one after the other, in order to gain time and to avoid ever talking of money." When he realized no money was coming for 1671 (the year had passed already), Graindorge became disgusted. He wrote a long letter to Huet summarizing the affair of the accounting and pouring

[37]Ibid., 606 (16 November 1671). As late as February 1672 Graindorge still could not account for how the money had actually been spent (ibid., 671 [11 February 1672]).

[38]Ibid., 606 (16 November 1671).

out his complaints. In the course of this letter Graindorge revealed that he had lied to Carcavi. Unable to report how Huet had spent the 1,000 livres in Paris, Graindorge altered the academy's accounts to omit certain items and to claim that only he (Graindorge) had spent anything, thus giving the impression that a large sum of money still remained. Furthermore, to cover up this deception (in case someone in Paris asked Huet to account for any money), he doctored his report to make it appear that he held everything that was left. Graindorge failed to specify precise amounts in this report to Huet, but it appears very likely that he had told Carcavi that a large part of the original 2,500 livres was still in his hands. In fact, Huet had less than 500 livres left, and Graindorge had nothing. Graindorge had expected the grant for 1671 to cover this deception until he could straighten out matters with Huet. Curiously, his actions amounted to the reverse of an embezzlement: he claimed to have spent less than he and Huet actually had. Since no money had appeared from Paris, however, Graindorge had to make up the difference from his own pocket. He wanted reimbursement from Huet.[39]

No evidence, either in Graindorge's letters or in Huet's account book, suggests Huet ever repaid Graindorge the money he advanced to hide the academy's financial problem. Indeed, rather than give back what little money he still held, Huet simply extended the academy's lease on his house into 1672 and paid himself the rent. The academy had Huet's house, but Graindorge was out a large sum of his own money. In effect, he provided the academy's financial "patronage" in 1671 and 1672, a situation that frustrated and dismayed him. Early in 1672 he decided to take the academy's future into his own hands.

THE ACADEMY GETS A CHAMPION

Despite the difficulties with the royal financing and his disgust with bureaucrats, Graindorge did not give up hope for the academy's success. In late January 1672 he conceived a new project—one intended to catch Colbert's attention and show him the value of the royal academy of sciences in Caen.

When a ship engaged in the New World trade put into port near Caen for refitting, the heavy covering of moss and barnacles on its

[39]Ibid., 672, 607, 671 (28 December 1671, 15 January and 11 February 1672).

bottom impressed everyone who saw the vessel. With the ship pulled ashore and careened for scraping, a species of Arctic sea duck known as the macreuse (scoter) began to feed on the shellfish covering the hull. The sight reminded the onlookers that the macreuse was not a normal bird, but generated spontaneously from the barnacles growing on the bottoms of ships. Soon reports on this marvelous sight made the rounds in Caen. Graindorge thought such a story was a silly tale. He decided to do a scientific investigation of the matter.[40]

Church practices, which treated the macreuse as a nonmeat food, gave him added incentive in this research. The faithful could eat the macreuse during Lent (presumably because of its spontaneous generation in the sea). But if, as Graindorge suspected, the macreuse actually reproduced through the same sexual processes that ordinary birds use, this ecclesiastical dispensation made no sense. Given this situation, Graindorge saw an opportunity to use his empirical science to expose a fallacy in accepted theological understanding of nature. In other words, he saw an opportunity to use his empirical science to fulfill a primary function of traditional natural philosophy: to help theologians apply their knowledge in correct formulations of religious practice. Accordingly, he planned a full-scale investigation of the anatomy and habits of the macreuse. In line with the rationale of traditional natural philosophy, such a project could become invaluable to the Church for establishing proper rules for fasting, but even more important from Graindorge's viewpoint, Colbert could not ignore such a work. The prospects pleased him:

> I shall be greatly relieved to know the reasons for the permission that has been granted for eating our macreuse during Lent. I was told that, when consulted, the Sorbonne declared that they are in no way flesh. I shall undertake a small discourse on this subject and anatomize several macreuses and other sea ducks to look at their eggs and the male genitalia, and I shall even have some paintings made. Perhaps this can be presented directly into M. Colbert's own hands without passing through those of M. Carcavi.[41]

Graindorge persuaded himself that this research could save the academy. He worried only that Pierre Carcavi and the Académie Royale might swallow it up before Colbert had a chance to see it. In mid-February, for example, he told Huet: "If I can put my work into a

[40]Ibid., 608 (29 January 1672).
[41]Ibid.

state to give to M. Colbert, your mediation could be nothing but pleasing and useful to our academy."[42] Although confident in the work itself, paranoid doubts about getting past bureaucratic barriers plagued Graindorge. To succeed, the treatise on the macreuse had to please Colbert. To please Colbert, Graindorge had to bypass the nitpicking "gentlemen" at the Académie Royale. Surely they had already sabotaged the academy in Caen. Graindorge had to find a way to stop them from doing so again.

For the next several months, reports on Graindorge's new knowledge of sea birds dominated his letters to Huet. Even finding himself alone at the academy failed to dampen his enthusiasm for this new work.[43] He worried only about finding a reliable intermediary who could present the finished treatise to Colbert. Late in April he produced his first draft of the work. Since Huet was unavailable, he asked Moisant de Brieux to become his editor. Given the views he had always held about the Grand Cheval, this was a notable turn of events. Graindorge's reentry into Brieux's sphere of influence suggests just how isolated he felt during 1672. He had lost contact with all the other academicians from the Académie de Physique.

Undoubtedly Brieux influenced Graindorge's treatise on sea birds. Perhaps Brieux even suggested the work.[44] Whatever his role in initiating the project, Brieux's influence soon shaped Graindorge's treatise. Initially the research had concentrated on the bird's anatomy and habits. Graindorge treated it as another *curiosité.* Yet under Brieux's influence, Graindorge took up the research of an *homme de lettres,* studying philology and the decisions of popes and councils.[45] The more Graindorge involved himself in such questions, the further he wandered from his own method of *curiosités.* Soon he seemed to forget every lesson from the Thévenot. As he did so, he stopped worrying about the other academicians. Graindorge planned to save the academy on his own.

When Graindorge cast himself as champion of the Académie de Physique, his confidence in organized science failed. The ideals of

[42]Ibid., 611 (25 April 1672).

[43]Ibid., 610 (1 April 1672).

[44]According to Graindorge's letters to Huet, he had not been attending the Grand Cheval regularly for several years before 1672. In January 1672, however, his letters once again begin to give reports on the activities of the Brieux group. The first discussion of the macreuse, in fact, appears in a report on discussions at the Grand Cheval (ibid., 608 [29 January 1672]).

[45]Ibid., 671 (11 February 1672).

group effort, *curiosités,* and unselfish contributions to knowledge all disappeared from his letters. His scientific vocabulary reverted to that of a natural philosopher and *homme de lettres.* The treatise on the macreuse led to further interest in shellfish, especially those spontaneously generated on ship bottoms. By early May he had completed a new treatise on this topic. As he described that work to Huet, he revealed his "new" attitudes. Although he conceived the work on sea birds as the "academy's" product, he saw the work on shellfish differently. It sprang from a part of Graindorge that was reemerging—the traditional natural philosopher. He wanted Colbert to know that this new work resulted from individual effort. In addition, in his account to Huet he cast the work strictly in terms that speak to the philologist's interest in the relationship between "things" and language:

> I have written a new work on the shellfish that our moderns call *concha anatifera* and our sailors call *satinette* since it is born from glazed planks [*des planches de satin*]. There is more reason to believe these shellfish are generated from the planks than that the macreuse is. . . . Since it is not necessary that this treatise serve only the academy, I asked M. l'Intendant to recommend me when he sends it to Colbert.[46]

At that point, Graindorge's devotion to his empiricism had ended. Soon Graindorge decided he should combine the treatises on shellfish and on the macreuse to produce one magisterial work. Surely it would regain Colbert's favor.[47]

During the summer and early fall, Graindorge labored to complete his new treatise. Finally he had it ready. Then, learning that Chamillart was going to Paris on business with Colbert, he entrusted it to the intendant and instructed him to put it directly into Colbert's hands.[48]

Chamillart did not return from Paris in mid-October 1672. Only then did Graindorge learn what the intendant had done with his treatise and an accompanying letter to Colbert. For Graindorge, the intendant's story sounded like a nightmare come true. Everything had gone wrong. According to Graindorge, Chamillart was nothing but a petty bureaucrat who bungled an important assignment. On 22 October 1672 Graindorge poured out his rage to Huet in a letter that marks the end of his commitment to the Académie de Physique:

[46]Ibid., 674 (9 May 1672).
[47]Ibid., 612 (20 May 1672).
[48]Ibid., 614 (18 September 1672).

> Having returned from his trip to Paris, M. l'Intendant told me this in passing: that he had put my work on the macreuse into the hands of M. Carcavi so he could deliver it into the hands of M. Colbert. [He said that he had done the same] with the letter I had sent to recommend the rest of the academy. [When he was himself with Colbert, Chamillart] did not even open his mouth. You judge by this business into what hands we have fallen. Between you and me, these gentlemen who are here such oracles are just little men like [me] in front of M. C[olbert]. . . .
>
> [Chamillart] only had to deliver a package and a letter and recover it at his discretion. There, we are more than ever under the power of M. Carcavi and under the discipline of the gentlemen of Paris.

In short, Graindorge's work never reached Colbert. Presumably the king's minister did not even know of its existence. Carcavi had treated Graindorge's work as a routine submission and handed it over to the academician Claude Perrault for evaluation. Graindorge had no confidence that it would receive a fair hearing under those circumstances. He knew all hope for renewing royal funding had ended: "Since a large part of my house is not being used, and since I was told that Madame Franquetot is going to rent your house, I took [the academy's] few pieces of equipment to my house. Only M. Postel and M. Chasles come there—looking very downcast. I shall not say a word to the others, who no doubt would not want to come to my house."[49]

In November Graindorge received the Académie Royale's official opinion on his treatise. The news came as one final bitter taste of royal science for Graindorge. The Académie Royale had not "approved" his work. Moreover, Perrault's complaints about it read like a condemnation of the kind of erudite natural philosophy that Graindorge had railed against for years: "I have heard from Paris that those gentlemen have found too many citations in my treatise on the macreuse. Also, that these gentlemen do not like to recall what other authors have said as much as they like to hear what is discovered for the first time."[50]

For Graindorge, the experience of being a royal academician had ended. There was nothing left to do but to vent his pent-up frustrations. Graindorge had always carefully phrased his letters to Paris, but at this point he saw little to lose in telling people what he thought. He wrote a letter to Gallois at the Académie Royale laying out all his

[49]Ibid., 616 (22 October 1672).
[50]Ibid., 630 (3 November 1672).

complaints against the other academicians in Caen, Carcavi, and the royal bureaucracy.[51]

Carcavi did not let Graindorge have the last word. When he learned about Graindorge's complaints, he announced publicly that the fate of the Académie de Physique involved just two issues: the academy had wasted its money on needless books and had not equipped the laboratories as it had promised. Within weeks Graindorge told Huet that Carcavi's remarks had reached the Hôtel de Ville in Caen: "Everything that M. Carcavi said . . . has been repeated here word for word."[52] With that public humiliation for Graindorge, the last recriminations over the fate of the royal academy of sciences in Caen ended. For a brief time Graindorge tried to keep the academy going as a private *assemblée,* but within months he gave up the effort.[53]

THE ACADEMY'S FAILURE

Any explanation for the Académie de Physique's closing must begin with royal financing. Graindorge could not account for all the money, nor could he justify those expenditures he did report. According to Carcavi, the academy had wasted too much money on books and had not equipped the laboratories. As a result, Colbert withheld further financing. In that sense, the Académie de Physique's failure was a simple matter. Inept handling of royal money destroyed the organization.

Fundamentally, however, these financial difficulties were an expression of a far deeper set of problems, the most important among them being a widely shared naiveté about what the academy needed in order to succeed. Those responsible for the academy—Graindorge, Huet, Chamillart, and ultimately Colbert—collectively possessed the resources the organization needed: an intellectual rationale, financial and material support, legal and political protection, and the social authority to direct activity. The academy's "patrons" failed to marshal these resources effectively.

The academy suffered a fatal weakness, a flaw that lay with the last

[51]Ibid.

[52]Ibid., 636 (30 November 1672).

[53]The last mention of any effort to hold sessions appears in Graindorge's letter to Huet from 10 February 1673 (ibid., 618): "Vous me demandez des nouvelles de notre Académie. Mr Postel el Mr Chasles sont les deux seuls que je vois."

of its patronage resources—social authority over the academicians. Huet and Chamillart possessed that valuable commodity in ample quantity, but time and again they proved unwilling (or unable) to exercise their power. All efforts to transfer their operational authority to Graindorge failed. In short, all concerned (except Graindorge) ignored the fundamental necessity of furnishing the organization with an appropriate *chef.* The fiasco that resulted from royal funding was just the most extreme example of their blindness. Given the academy's record between 1665 and 1670, Huet and Chamillart must have foreseen problems in allowing Graindorge to administer royal funds. Nevertheless, they did so.

This is the knottiest fact to emerge from the Académie de Physique's history. Everyone from the academicians to Colbert seems to have thought that once the academy had funding, it could quickly establish the momentum necessary for long-term success. Unfortunately, the Académie de Physique needed more than funding. Finding a reliable *chef* was more important than a sophisticated research program, royal incorporation, or royal funding. Without a *chef,* the royal academy ceased to function, just as the private-patronage society had ceased to function during Huet's extended absence in 1666. Graindorge recognized this fact, and his letters to Huet indicate that his patron did also. Yet neither Huet nor Chamillart took effective action to deal with the issue. One can only speculate about their motives, but two possibilities warrant consideration. First, these "patrons" may not have understood the full implications of Graindorge's need for continuity at the sessions. Chamillart certainly acted as if he thought the academy's work was simpler to complete than it was. Second, responding to his own social imperatives, either Huet or Chamillart could have claimed it was beneath his dignity to deal with the academy's day-to-day operation. The vanity of his social position could have convinced either of them that the problem was minor.

Graindorge referred to Huet as the academy's *fondateur* and to Chamillart as its *instituteur.* In doing so, he showed perceptive understanding of the organization's needs. He looked to these men to animate the group. He did not want their scientific expertise; he needed them as "patrons" (in the sense implying *un grand personnage*) whose presence would bring the academicians to the sessions and whose authority would settle disputes. He needed them to keep the academy together. Graindorge recognized he could not fill the role of *chef.* He also recognized that the presence of one of these men constituted the first requirement for success in any collective scientific program.

Graindorge's inability to provide effective leadership and his mismanagement of royal funds led directly to the closing of the Académie de Physique. Nevertheless, it would be unfair to conclude that his inadequacies caused the academy to fail. More than anyone else, he recognized his own limitations and tried to deal with them. More important, he faced substantial obstacles. As attractive as the idea of a cooperative, empirical program seemed, it was definitely a novelty in seventeenth-century France. Practitioners of this new science appreciated the concept, but it did not reflect contemporary French social values. Graindorge's "scientific academy" fitted poorly in a world that valued hierarchy, status, and the role of patronage.

Graindorge astutely recognized what he needed in order to establish the Académie de Physique as a viable institution. More than either Huet or Chamillart, he knew the first priority had to be a shift in values that would bring the individual academicians together in support of a new form of activity. Under the circumstances he faced, his most practical approach to that goal—establishing an academy—lay in using the accepted social norms related to patronage. He had to use this system to build a consensus on the importance of the academy's work. Only then could he hope to free the organization from its dependence on Huet and Chamillart.

Nor was Graindorge alone in facing this problem. Of the half-dozen or so scientific patronage groups most active in France during the half century before the academy's demise, not a single one survived intact following the founding patron's withdrawal from active participation.[54] Individuals, or even small groups, often migrated from disintegrating organizations to new patronage circles (as happened, for example, with the closing of the Montmor in 1663), but the most salient characteristic of all such reformulations was the lack of continuity in organizational mission, programs, and personnel. Even the new Académie Royale felt these effects, both in its founding and in its subsequent history. Many historians have noted the diffi-

[54]For a general survey of such groups as the Peiresc circle in Aix, Mersenne's following, the cabinet of the brothers Du Puy, the Montmor and Thévenot academies, and the Boudelot and Justel circles, see Brown, *Scientific Organizations*. For more particular information on individual groups, see Solomon, *Public Welfare, Science, and Propaganda;* for the Renaudot academy and for Parisian patronage in the 1650s, see Jean Mesnard, "Pascal à l'Académie le Pailleur," *RHS* 16 (1963): 1–10.

culty in tracing the Académie Royale's organizational antecedents; likewise, many have remarked on its seeming decline in the period between Colbert's death in 1683 and the reorganization in 1669. As the most perceptive account of the Académie Royale's history asserts: "While Colbert lived, the Academy flourished."[55]

Graindorge's problem was unique only in degree, not in kind. He simply faced a more acute problem than did other potential patrons or *chefs* who contemplated formulating a new organization following the disintegration of an old one—for two reasons. First, Huet, not Graindorge, had recruited the academy's membership. Second, the royal incorporation had officially fixed that membership. Three of the members—Villons, Lasson, and Savary—clearly held social positions well above that of Graindorge's, and he never had any real chance of gaining their support for his leadership. With those at approximately his social level in the recently ennobled minor nobility—La Ducquerie, Vaucouleurs, and Cally—he fared well, and he even counted on Vaucouleurs's support. Yet it was with those who were clearly his social inferiors that Graindorge fared best. Among that group—Hauton, Chasles, Busnel, Postel, Daleau, and Vavasseur—his only real problem lay with Daleau, who had gained his spot as technician despite Graindorge's repeated warnings about the man.[56]

With the membership, the organization's form, and the program frozen after royal incorporation, Graindorge had little chance of galvanizing the Académie de Physique. And with Huet and Chamillart as absentee "patrons," his position became untenable. In any other circumstances, Huet's final withdrawal from the scene in 1670 would have allowed Graindorge to reformulate a new patronage *assemblée* at his own house. He had the nucleus for such a group in Hauton, Chasles, Vaucouleurs, and Postel. Indeed, after he had removed the academy's equipment from Huet's house in October 1672,

[55]Hahn, *Anatomy of a Scientific Organization,* p. 17. Hahn is less convinced than most of the severity of the Académie Royale's decline between 1683 and 1699 (pp. 19–21). Nevertheless, even while calling for further study of the academy's development over that period, Hahn recognizes that Colbert's death brought distinct changes to the organization.

[56]See, for example, Graindorge's comments in his letter to Huet on 14 May [1667]: "Je ne crois rien Daleau qui fait le mysteriux a la reputation de savoir quelque chose. Lon me demanda sil nestoit point de notre Academie physique. Je dis quil estoit trop particulier quand je dirois escroc. Je crois que je ne me tromperois car il a eu 30 pistoles dune personne qui na rien appris de luy." Obviously, since Graindorge was describing the man for Huet, Daleau did not yet belong to the academy at that point.

some of these men still continued to appear at his doorstep on Thursday evenings. He may even have been able to recruit his friend Lasson to participate in this less fractious group.[57] Under the circumstances created in the royal incorporation, however, Graindorge had little chance to formulate such a circle. After late 1672 he showed no enthusiasm for such a project.

But then, this was the same problem Graindorge had faced from the earliest days following his return from Paris. As we saw in Chapter 1, Huet and Graindorge (as patron and *fidèle*) formed an almost perfect combination in creating a scientific research institute. Huet possessed the *état* necessary to recruit in the Republic of Letters (thus raising the status of their activity) and Graindorge possessed a vision constituting a new form of science. Whenever they found themselves together in Caen, the academy flourished. Whenever Huet removed himself, Graindorge found it impossible to wield effective control. Graindorge's ability to guide the academy in August and September 1668 was an aberration, his performance at other times the norm. Indeed, in guiding academicians whom his patron Huet had recruited, he fared no better nor worse than we might expect in any established patronage group whose patron tried to transfer loyalty wholesale to a new *chef*.

If we must blame Graindorge for any failure, it should be for his inability to galvanize his patrons—not for his inability to motivate the academicians or for his financial mismanagement. He never persuaded Huet and Chamillart to give him the sustained period of attendance necessary to instill the academicians with a sense of purpose in doing scientific or technological research. Ultimately, then, the only satisfactory explanation for the failure of the Académie de Physique must lie in the nature of the relationship between the individual academicians and their patrons: Graindorge was unable to break the ties that bound them together. In the end, he found it impossible even to break his own bonds. After all, when royal funding created the choice between pleasing his patron and meeting the academy's larger needs, he followed the dictates of patronage. Therein lies the explanation for the academy's demise.

[57]The best evidence on this point comes from Graindorge's response to one of Huet's queries on 10 February 1673 (BL, A 1866, 618). Clearly, even though the academy no longer met, he was still in touch with the academicians Postel, Chasles, Vaucouleurs, Lasson, and Vavasseur and knew what their interests were.

7
Royal Administration, Patronage, and Science

Throughout his letters to Huet, André Graindorge offered sharply drawn vignettes revealing his attitudes toward almost every individual involved with the Académie de Physique. In a few cases he expressed genuine friendship. When he was separated from Huet, for example, Graindorge wanted nothing more than for the two of them to be together again—to sit by a fire drinking warm cider and playing chess. More often he offered less kindly portrayals: Chamillart was full of promises but had feet of clay; Pierre Carcavi proved himself to be an untrustworthy "scavenger"; Daleau was a "swindler" and "cheat." In only one case did Graindorge fail to give Huet a sharp verbal image defining his attitude toward someone who played a role in the academy's life: he never revealed any strong feelings toward Jean-Baptiste Colbert. In Graindorge's letters, Colbert always remains just offstage, ready to enter the scene but never given his cue—a curious phenomenon, and one we must address. It is an issue that brings the Académie de Physique's historical significance into sharp focus.

We have already seen a great deal of evidence on how André Graindorge misinterpreted the Académie de Physique's royal incorporation. He wanted a research academy; he sought patronage to produce it. When he encountered Chamillart, the intendant's obvious desire to use the Académie de Physique's research capability led Graindorge to think he had found a new patron. He was wrong. Chamillart also wanted a research academy, but he offered bureaucratic administration—not patronage—to produce it. The difference in their perceptions proved fatal to the academy they both wanted.

This clash between two perceptions of the academy's organizational form lay at the heart of the events that shaped it for five years—from late 1667 through 1672. The two forms proved to be incompatible. To arrive at any final statement on the academy's history, we need to know why. Was this simply a local problem between Graindorge and Chamillart? Or were their radically different conceptions of organizational form symptomatic of what might happen when any private-patronage organization became subordinated to royal service? With the Académie de Physique, the means to address this issue is through Colbert's role.

Ultimately Colbert was the minister responsible for the Académie de Physique. The question of whether he took a direct role in administering the organization is central to its history. Yet Graindorge's letters appear to offer contradictory evidence on this point. Was Graindorge correct when he asserted that Colbert asked to read the reports from Caen? Or was he right when he insisted that Chamillart and Carcavi withheld information from Colbert? These questions are probably unanswerable in such specific form, but with more general phrasing they become keys to understanding the academy: To what degree did Colbert concern himself directly with the Académie de Physique?

COLBERT'S ROLE

In the context of Colbert's ministry, the contradictions in Graindorge's letters become comprehensible. It appears that Graindorge was wrong when he insisted that Chamillart and Carcavi kept Colbert ignorant. Given Colbert's administrative style and the record of his specific dealings with both Guy Chamillart and Pierre Carcavi, it becomes difficult to accept Graindorge's contention that Colbert lacked knowledge of the way his *fidèles* dealt with the academy in Caen.[1] In

[1]More than 150 letters from Chamillart's correspondence with Colbert have survived (BN, Mélanges de Colbert, 102–170). These letters, spanning the period from 1661 to 1675, are obviously far from complete. During critical periods of the Académie de Physique's history the documentary record is fragmentary at best. For instance, only three of Chamillart's letters to Colbert from August 1667 through March 1668 have survived (ibid., 147, ff. 84, 597, and 609). None of those letters mentions the Académie

fact, the opposite appears much more likely. Despite Graindorge's suspicions, Chamillart and Carcavi almost certainly told him the truth when they claimed to act on Colbert's personal instructions.

To André Graindorge, Colbert was an aloof, distant, and powerful figure. Frustrated and bitter, Graindorge came to see Colbert as his last hope for reviving the academy. It is clear that as Graindorge invested more and more energy in the treatise on the macreuse, he began to idealize Colbert into almost mythological proportions. To Graindorge Colbert became the wise and all-powerful patron—a "Little Father"—surrounded by conniving minions. According to Graindorge, these sycophants kept Colbert ignorant and thus prevented him from dispensing justice. If he could only get past those pettifogging bureaucrats Chamillart and Carcavi, Graindorge was sure the academy's true patron—Colbert—could set everything right.

Graindorge's idealized version of Colbert's role as patron is difficult to reconcile with the standard accounts of how the minister functioned.[2] Among historians of the seventeenth century, Colbert is famous (or notorious) for his meticulous attention to detail.[3] Moreover, he was capable of lavishing attention on the academies he created.[4] The Academy of Inscriptions met at his house; he wanted regular progress reports and planning documents from the Aca-

---

de Physique or any plans for such an organization. Likewise, for the period from July 1669 through August 1673 (ibid., 154–158), only eight of Chamillart's letters to Colbert have survived. On Colbert's dealings with Carcavi, see *Colbert, 1619–1683,* p. 379; Murat, *Colbert,* p. 180.

[2]It is unlikely that Graindorge ever met Colbert. He did know a great deal about the minister, however. While Graindorge attended the Thévenot, for example, he reported that Colbert gave various members of that organization specific commissions to investigate scientific questions. The best example of Colbert's dealings with the Thévenot involves a petitioner who sought a royal *gratification* for squaring the circle. Before Colbert would allow Pierre Carcavi to give a *gratification* of 200 pistoles to this "squarer of the circle," he commissioned the mathematician Gilles Personne Roberval to evaluate the man's claims. Roberval brought the problem to the Thévenot. According to Graindorge, this "schemer's" proof hoodwinked Roberval and all the other mathematicians at the Thévenot until Adrien Auzout finally "disabused" everyone of their errors in accepting the fraud. Graindorge's account of the episode is found in his letters of 16 September and 23 September 1665 (VDLI, pp. 319–322; BL, A 1866, 1983).

[3]For historiographic and bibliographic discussion on how Colbert's ministry has been interpreted, see William F. Church, *Louis XIV in Historical Thought* (New York, 1976). For more recent discussion and bibliography, see Murat, *Colbert;* Meyer, *Colbert;* and *Colbert, 1619–1683.*

[4]Charles Perrault's *Mémoires de ma vie* remains the best source on Colbert's relations with royal academies. See also Murat, *Colbert,* pp. 173–192, 381–387.

démie Royale des Sciences; and he barraged Charles Errard at the Académie de France de Rome with detailed instructions on how to direct that organization.[5] In short, Colbert showed a high degree of interest in the king's academies. If he ever considered the Académie de Physique a legitimate royal academy, it becomes impossible to accept Graindorge's perception of this minister as an aloof and distant figure.

While castigating Chamillart and Carcavi, Graindorge idealized Colbert. Simply by calling Graindorge's perceptions of Chamillart and Carcavi into question, however, his letters begin to reveal their own evidence on Colbert's involvement with the Académie de Physique. In late 1671, for example, as Graindorge tried to account for the academy's royal money, Carcavi told him that Colbert wanted an annual summary of the organization's work and expenses—both to ensure that the money was put to good use and so Colbert could increase the budget as soon as it became necessary. Graindorge came to see these kinds of statements from Carcavi as officious meddling and as a form of bureaucratic procrastination, but he was surely wrong.[6]

Graindorge also suspected that Carcavi indulged in a deception when he claimed Colbert had taken a personal interest in Lasson's skill at constructing telescopes. Almost certainly Graindorge was wrong once again. According to Carcavi, Colbert offered to furnish all the materials needed plus a substantial *gratification* if Lasson would just try his hand at constructing a large telescope. Nothing ever came of this offer—not because Colbert or Carcavi reneged on the promise but more likely because Graindorge never told Lasson about it. On such issues, nothing but Graindorge's growing distrust for royal officials casts doubt on Carcavi's claim that Colbert would "effect everything he has promised."[7]

Upon careful reading, Graindorge's letters reveal critical evidence tying Colbert to the Académie de Physique. In fact, if we discount the florid rhetoric Graindorge used to vilify those royal officials he wanted to blame for destroying the academy, his own letters reveal a

[5]Perrault, *Mémoires*, pp. 37–42; Hahn, *Anatomy of a Scientific Institution*, p. 17; George, "Genesis," pp. 372–401. Numerous examples of Colbert's directives for Errard can be found among the documents Pierre Clément edited for *LIMC*, vol. 5.

[6]BL, A 1866, 670, 607 ([late 1671], 15 January 1672). Beginning at the end of 1667, Colbert had made exactly the same demand on the Académie Royale (George, "Genesis," p. 398).

[7]BL, A 1866, 670 [late 1671].

Colbert who kept himself informed and current on the state of the organization; for instance, Colbert was probably involved even at the level of ordering the Académie de Physique's dissection reports included in a volume on natural history (just as Carcavi claimed).[8] Certainly the Colbert who emerges upon such a close reading of Graindorge's letters bears a closer resemblance to the minister modern historians portray than does the idealized patron Graindorge preferred to see.

To see Colbert through Graindorge's letters, then, requires us to discount vivid characteristics of Carcavi and Chamillart. Consistently, from 1667 through 1672, Graindorge misunderstood what royal officials wanted—and what they were doing. This was certainly the case with Chamillart and Carcavi, but his understanding was no more distorted in those cases than in that of Colbert. Colbert was not the stylized patron Graindorge sought. Like his *fidèle* Chamillart, the minister Colbert wanted new technological and scientific products from the Académie de Physique. Aside from his offer to support Lasson, which ultimately came to nothing, his desires in this regard are amply demonstrated by his willingness to reimburse the extraordinary sum of 1,200 livres that the monk claimed as expenses for his trip to Paris, in the 1,200-livre *gratification* that Hauton earned for his desalinization project, and in the royal appointment that followed Villons's presentation of his chronometer in Paris.[9] Obviously Colbert not only wanted such things but was willing to reward anyone who produced them, even when the project failed to give any satisfactory solution to the technological problem it addressed.

Merely expanding the context of the Graindorge letters, then, exposes a Colbert who was directly involved with the Académie de Physique. Moreover, he was willing, even eager, to support the organization's scientific and technological work. In that sense the Graindorge material goes a long way toward establishing Colbert's impact on the academy. Nevertheless, this line of inquiry cannot entirely explain Colbert's part in the organization's history. More particularly, tying Colbert directly to the Académie de Physique gives two entirely new twists to the question about his role: Why did he cut off funding after 1670? And could Colbert have become the academy's personal "patron"? Once again, given the available evidence, these questions

[8]Ibid., 603 (13 November 1670).

[9]Chamillart to Colbert, 28 February and 25 March 1669, BN, Mélanges de Colbert, 149, ff. 661r–662r, 949r–v; J. J. Guiffrey, *Comptes des Bâtiments du Roi sous le règne de Louis XIV*, vol. 1 (Paris, 1891), p. 388; BL, A 1866, 655 [late 1670].

are overly specific. Phrased in a more generalized form, however, they become answerable: Have we any evidence of clashes between the ideals of private patronage and the bureaucratic imperatives Colbert followed in his dealings with the academy?

ROYAL ADMINISTRATION VERSUS PATRONAGE

The most suggestive evidence on the imperatives governing Colbert's relations with the Académie de Physique is found in one of his own letters. In mid-1673 Colbert rejected a request from the duc de Saint-Aignan, the governor of Normandy, who wanted to replace Colbert as "protector" (patron) to a royal academy. Indeed, in refusing Saint-Aignan, Colbert issued a stinging rebuke: "I find it embarrassing to me when a duke and peer and First Gentleman of the Chamber asks me to become the associate protector [*sous-protecteur*] to an academy for which the king has already ordered me to become the protector. I certainly know you yourself would be embarrassed."[10] Colbert did not specify the academy Saint-Aignan sought to "protect." For the moment, however, that ambiguity matters less than what the reprimand indicates about Colbert's attitude toward the academy in question: he was jealous of his rights over the organization.

Tracking down which royal academy Saint-Aignan might have become involved with during 1673 reveals something extraordinary. Three pieces of evidence indicate that Saint-Aignan wanted to replace Colbert as "protector" to the Académie de Physique. First, of the seven royal academies Colbert's letter might concern,[11] the most logical is the Académie de Physique in Caen. This was the only academy Colbert "protected" that might have interested Saint-Aignan.[12] Second, Pierre-Daniel Huet claimed that Saint-Aignan "urgently" sought to

[10]*LIMC*, 5:350 (17 July 1673).

[11]The Académie Française (1635), the Académie Royale de Peinture et Sculpture (1648), the Académie des Inscriptions et Belles-Lettres (1663), the Académie Royale des Sciences (1666), the Académie de Physique de Caen (1667), the Académie d'Architecture (1671), and the Académie de Musique (1672).

[12]Intellectually the Académie Française is the only royal academy that might have interested Saint-Aignan. This academy cannot be in question here, however, since Louis XIV himself (not Colbert) acted as its *protecteur*. In fact, when one restricts the list of possible academies to those for which Colbert was officially recognized as *protecteur* (the Académie des Inscriptions, the Académie Royale, and the Académie de Physique), it becomes almost inconceivable that Saint-Aignan's attempt to demote Colbert to the role of *sous-protecteur* could have concerned any academy other than the one in Caen.

participate in the Académie de Physique.[13] Third, the date of Colbert's letter (17 July 1673) fits all requirements for tying this incident to the Académie de Physique. Taken together with the fact that Saint-Aignan was the governor of Normandy, these three pieces of evidence present a strong case for the Académie de Physique as the royal organization in question. Before the significance of that case can be evaluated, however, the issue of dating needs clarification. If the Académie de Physique ceased to exist in October 1672, as Graindorge claimed, how could Colbert refuse to let Saint-Aignan become its patron in July 1673? To unravel this conundrum we must take a closer look at the issues involved in the academy's closing. Graindorge's letters noted the cessation of all scientific activity, but he did not have the authority to terminate the existence of a royal corporation. And if he did not, who did?

As we saw in Chapter 6, Graindorge considered the Académie de Physique dead in late 1672. When he learned that Chamillart had given the treatise on the macreuse to Carcavi, he stopped all efforts to revive the academy. In the same letter that reports how Chamillart had failed him, for example, Graindorge told Huet he was going to abandon the academy and turn his attention to new projects.[14] He followed through on that pledge. Over the next three years his letters offer nothing to suggest he made any effort to resuscitate the Académie de Physique. Chasles and Postel came to his house occasionally on Thursday evenings, but Graindorge described their activity as a discussion group or *assemblée.* As far as he was concerned, the academy had died in October 1672.

Graindorge's firm conviction that the Académie de Physique had ceased to exist did not put any kind of official stamp on the organization's demise, however. He said nothing to indicate that anyone in

[13]Huet made this claim twice in the *Commentarius* (pp. 229, 317). The passage referred to here reads: "Tam feliciter autem increbuit florescentis hujus Academiae fama, ut Bellovillarius, Dux Santanianus, literariae gloriae percupidus, in eam optaverit admitti, suumque nomen ut in Academicorum nostrorum seriem referretur valde à me contenderit" (p. 229). By tying this episode to the duke's desire for "glory" in the Republic of Letters, Huet's comment suggests that Saint-Aignan had offered patronage.

[14]Graindorge told Huet that he planned to begin work on a translation of Lucretius. Of equal importance, he also stated he was not going to tell the academicians that he had removed the academy's equipment from Huet's house (BL, A 1866, 616 [22 October 1762]). In effect, he claimed he would do nothing to revive the Académie de Physique. Graindorge's commitment to his translation is also borne out by a letter Moisant de Brieux addressed to Henry Oldenburg on 5 June 1673 (*CHO,* 9:670–673). Brieux refers Oldenburg to Graindorge for news of scientific activity in Caen but also mentions that Graindorge was working on his translation.

Paris ordered him to close the academy, or that the Parisians already considered the organization dead. Indeed, no surviving document suggests that any royal official ever recognized the Académie de Physique's demise. Carcavi had only told him that Colbert was not renewing the royal funding for 1671 (or 1672) unless Graindorge could justify the expenses for 1670. But Carcavi had relayed Colbert's ultimatum in late 1671—not in late 1672. Even Graindorge thought the "academy" continued to exist for another year after the ultimatum on funding. Since the Académie de Physique had enjoyed royal funding during only one of its five years as a royal organization, it is unlikely that anyone in Paris ever considered the issue of money as critical to its existence.

The situation was much the same in Caen. Graindorge's letters offer nothing to suggest that anyone other than himself had reason to date the academy's demise to October 1672. His own letters reveal that the last time any appreciable number of academicians had come to sessions was at the end of 1670—in the weeks just after Chamillart ordered the group to return to Huet's house. In short, Graindorge saw Chamillart's handling of the treatise on the macreuse as killing the Académie de Physique, but nothing suggests that anyone agreed with him. Huet certainly did not consider the academy beyond all hope in late 1672. Once again, Graindorge's own letters offer the essential evidence on this point: they reveal that Huet asked for news of the academy several times in the months following October 1672.[15]

Graindorge's dating of the academy's failure was personal. In late 1672 *he* stopped trying to revive the organization. Most of the academicians had stopped work almost two years earlier; Huet had not given up hope for a revival months later. For Chamillart, Carcavi, and Colbert, nothing changed in October 1672. Why should Graindorge's disappointment over what happened to the macreuse treatise make any difference to anyone? Graindorge himself knew that Chamillart had not invested the events surrounding the delivery of the treatise with any particular significance.[16] Put in that context, Graindorge's dating of the academy's demise was idiosyncratic.

[15]The clearest demonstration that Huet thought the academy still in existence appears in Graindorge's letter of 10 February 1673 (BL, A 1866, 618): "Vous me demandez des nouvelles de notre Académie." He then proceeds to give Huet news about the academicians Postel, Chasles, Vaucouleurs, Lasson, and Vavasseur. The content of his remarks makes it clear that the academy was not meeting at this point.

[16]This conclusion is based on the way Graindorge introduced the subject for Huet in his letter of 22 October 1672 (BL, A 1866, 616). The intendant had told him "in passing" what he had done with the treatise. In his interview with the intendant,

Graindorge's assertions about when the academy died are significant for just one reason. He had been the moving force behind the idea of a sustained, cooperative research program. Once he gave up on that idea, no one else associated with the academy was likely to push for its revival. Thus his recognition of defeat marked the end of all possibility that the academy might establish itself as an organization devoted to the pursuit of *curiosités* and *expériences*. But acceptance of the significance of Graindorge's dating does not imply that any of his contemporaries knew about it, much less that they agreed with it. With that established, an evaluation of the full significance of Colbert's letter to Saint-Aignan becomes possible.

For any period before October 1672 it is impossible to reconcile Huet's assertions about Saint-Aignan with the material in Graindorge's letters. Discussion of Saint-Aignan's name in association with natural philosophy and science never appeared in those letters.[17] Thus if we accept Huet's account in the *Commentarius* as reflecting an actual event, as it surely does, Saint-Aignan's attempt to find glory as patron to the Académie de Physique must have occurred later. Under the circumstances, the only logical time for Saint-Aignan's involvement was in the early months of 1673, during the period when Huet was still asking Graindorge for news of the academy and its academicians. In that case, the timing of Colbert's letter becomes comprehensible.

If Colbert's letter to Saint-Aignan involves the Académie de Physique, there is only one way to describe his role in the organization's history. It was Colbert, not Chamillart or Carcavi, who determined the academy's fate. Colbert told Saint-Aignan that he found it "embarrassing" when a duke and peer asked him to step down from a role the king had "ordered" him to "perform." In effect, those words

---

Graindorge said nothing to indicate how important he thought the event was: "Je nen temoigné pas mon chagrin parce que dans le temps ou nous sommes nous avons besoin de tout mais Jen ay esté bien mortifié."

[17]Huet himself did not meet Saint-Aignan until sometime after 1667 (Tolmer, *Huet*, p. 368; *Commentarius*, p. 229). Despite Tolmer's claim that Huet met Saint-Aignan for the first time during the Christmas season of 1667, their first introduction must have occurred later. Tolmer places Saint-Aignan (and his meeting with Huet) in Caen; yet Huet was absent from Caen from mid-October 1667 to mid-March 1668. Saint-Aignan's name appears only once in any of Graindorge's letters—in that of 16 January 1668—and the reference concerns Saint-Aignan's interests in belles lettres rather than science (BL, A 1866, 572). Given Huet's extensive absences from Caen following 1666, it is almost inconceivable that Graindorge could have avoided all mention of Saint-Aignan if the duke had had any interest in the academy.

placed the organization under a royal mortmain; they ended all hope for any revival through private patronage. Once the academy belonged to the monarchy, it could never return to private patronage.

Given the circumstances in Caen, Colbert's letter amounted to a death sentence. But this was a death sentence of a very peculiar kind. The academy was already in total disarray: it was leaderless, it was unfunded, and it had nowhere to meet. Indeed, it had not met regularly since 1670. By any conceivable criterion, it was already dead. Its only hope for resurrection was the same as it had always been: finding an acceptable replacement for Huet. Denial of Saint-Aignan the opportunity to become the academy's patron or protector in 1673 amounted to withholding of life-support systems from a trauma victim. Colbert did not kill the academy, but his letter to Saint-Aignan determined its fate.

To modern eyes, Colbert's rebuff to Saint-Aignan may appear paradoxical: protecting his rights over a royal academy, he denied that academy any chance to survive. Within the terms of the Académie de Physique's history, however, his action becomes entirely comprehensible. He simply treated the academy the same way Chamillart had always done. As we saw in Chapter 4, Chamillart ostensibly wanted the academy's technical services, but in practice he gave that desire lower priority than the imperatives for establishing a royal corporation. Chamillart proved himself scrupulous in carrying out the administrative requirements for tying the academy to Paris but slipshod in ensuring that the organization mounted a coherent research program.

Although Colbert desired technological and scientific products, it is clear that for him, too, establishing the corporate form of the royal academy counted more than the work produced. Undoubtedly Chamillart's day-to-day practices as *chef* to the Académie de Physique actually represented the extension and implementation of Colbert's administrative dictates. Chamillart's behavior certainly indicated that he thought he was carrying out Colbert's instructions. Their shared attitude, in fact, is one in which the academy's existence broke down into two components. The first involved the actual research program (Colbert and Chamillart wanted the academy's scientific and technological products). The second involved the legal fiction embodied in a royal corporation (they wanted that royal corporate personality to exist). Colbert tried to help establish the first component (the research program), but even after that effort failed, he insisted that the second (the royal corporation) still lived. The king had created a royal academy of sciences in Caen. Moreover, Louis XIV had ordered Colbert to "protect" it. Legally, therefore, the king owned the academy; Colbert "protected" it—even if it never held another session.

Despite the idealized notions in Graindorge's letters, the Colbert who "protected" the academy was a royal minister, not a patron. Stripped of his cynical portrayals of Chamillart and Carcavi, Graindorge's own letters yield the evidence that Colbert instructed these officials directly in all their dealings with the Académie de Physique. Likewise, to question Graindorge's romanticized version of Colbert as "patron" is to reveal a royal minister following the imperative common to all royal officials in Louis XIV's France: Colbert guarded the monarchy's interest first. Given that imperative, what can we say about its influence on the Académie de Physique's science? We can now turn to this one last question.

THE ACADEMY'S SCIENCE

During the five years that Graindorge accepted as constituting the Académie de Physique's life as a royal institution (1668–1672), the organization produced (or sponsored) at least six submissions at the Académie Royale des Sciences. Almost certainly, however, the actual number of submissions amounted to eight projects. Any evaluation of the scientific program at the academy must be based on those eight projects. Colbert had to use their quality to determine the academy's value. We cannot do less in discussing its historical significance.

Given the classification of knowledge that operated for the Académie de Physique, its eight submissions in Paris represented a grab bag of topics, including natural history, astronomy, chemistry, natural philosophy, and horology. For all practical purposes, however, these eight projects fell into two clear-cut categories: four were scientific treatises in vertebrate anatomy; four were technological projects. When we search Graindorge's letters for evidence to judge the success of these submissions to Paris, a now-familiar pattern emerges. Graindorge often expressed strong opinions on their value, yet the very way he described them for Huet indicates that others held equally strong views that were very different from his own.

Direct evidence shows three submissions in anatomy to the Académie Royale des Sciences, and strong circumstantial evidence indicates a fourth. The dissection report on the partially rotted sturgeon (June 1668) was the Académie de Physique's first scientific submission to the Académie Royale. This project was definitely a success in Paris. The academy's secretary, Jean Gallois, read the entire treatise to the assembled company, recorded its contents verbatim in the *procès verbaux*, and then relayed the Académie Royale's compliments to Caen.[18] For the 1660s this was a significant piece of anatomical work.[19] The only way to evaluate this project is as the Parisians did: as proof that the Académie de Physique had significant research potential.

The academy's next two anatomy submissions went quite differently. It is impossible to trace the progress of these works through the Académie Royale. Indeed, only indirect evidence shows that the first of them must have gone to Paris. In describing these submissions, Graindorge's letters identify them only as concerning marine anatomy, with each consisting of a series of dissection reports accompanied by Du Fresne's drawings. The first of these collections was ready to send to Paris in September 1668, the second in September 1670. Beyond that, we can say nothing specific about their content or even about the number of individual reports they contained. The key issue concerning these works, however, involves Graindorge's attitude toward them. He described the collected dissection reports as "completed" scientific works. In his mind, their completion gave them special status.

In the 1668 case, Graindorge definitely thought that the material the academy had produced constituted a coherent manuscript, one the academy could publish as a commercial venture. Ultimately nothing came of that plan. Although it is impossible to know exactly why, the events surrounding the Académie de Physique's next major anatomy submission suggests an answer. In the 1670 case, the Académie Royale's reaction to the quality of the dissection reports from Caen was extremely positive—so much so, in fact, that the Parisians wanted to integrate the individual reports from Caen into a treatise on natural history as Colbert had ordered. Understandably, Graindorge reacted with bitterness and cynicism. Sending the Académie de Physique's work to Paris had robbed his academy of the honor of publishing it.[20]

[18]AdS, Reg (Physique), vol. 4, f. 116v (21 July 1668).
[19]Tolmer, *Huet*, pp. 376–378.
[20]BL, A 1866, 576, 603 (24 September 1668, 13 November 1670).

Graindorge's own treatise on the macreuse became the fourth and last anatomy project submitted to the Académie Royale des Sciences. Once again Graindorge had a strong negative reaction to what happened in Paris. Indeed, his experience with this treatise crystallized his understanding of the relationship between Caen and Paris. When he saw how Chamillart, Carcavi, Gallois, and Claude Perrault treated his work, Graindorge finally realized that any scientific products created at the Académie de Physique fell under royal ownership. A full two years after he had received the Académie Royale's negative evaluation of his treatise, Graindorge had not been able to recover the manuscript or the illustrations he had sent with it. At that point (January 1674), as he commented on his efforts to recover the macreuse treatise, he summarized exactly what had happened with every anatomy submission from the Académie de Physique: "Anything entering the coffers at the academy in Paris never reappears."[21]

With each anatomy submission, the same pattern prevailed. Graindorge saw these works as publications that should reflect honor and glory on the Académie de Physique; the Parisians treated them as interesting and valuable technical reports from a subordinate academy. Whatever the Académie Royale's final evaluations may have been, once the works arrived in Paris, they no longer belonged to the Académie de Physique. Given the demands Chamillart, Carcavi, and Colbert made concerning the academy's ability to produce something—almost anything—to justify its existence, Graindorge came to think that the Académie de Physique's status as a royal academy amounted to a license for the Parisians to rob his provincial organization of all its best efforts. He had reason to believe that the Académie de Physique lived "under the discipline of the gentlemen in Paris."[22]

Despite Claude Perrault's negative comments about his treatise on the macreuse, Graindorge continued to have faith in the work's scientific merit. Apparently others agreed, because Thomas Malouin, a member of the Faculty of Medicine at the university in Caen, finally managed to publish it in 1680–eight years after Chamillart carried it to Paris, four years after Graindorge died.[23] As with publication of the 1668 dissection report on the sturgeon (which finally reappeared to the world in 1733 as part of Du Hamel's *Regiae Scientiarum Academiae Historia*), its appearance came too late to do any good for Grain-

[21]Ibid., 624 (20 January 1674).
[22]Ibid., 616 (22 October 1672).
[23]Brown, "L'Académie de Physique de Caen," p. 121.

dorge or for the Académie de Physique.[24] In effect, just as Colbert had placed the academy's corporate personality under a royal mortmain, his requirement to submit work to Paris had done the same for its scientific program.

Of course, Graindorge's perspective on these matters was not the only one that counted. The sole evidence available on the Parisian position, however, comes through Graindorge's letters. Such evidence is thus heavily shaded against the Parisians. Nevertheless, enough of the Parisian position comes through to permit us to see the rationale behind their actions. Whereas Graindorge's attitudes followed the dictates of the patronage system governing the Republic of Letters, the Parisians behaved as if the Académie de Physique was just one cog in a bureaucratically administered research machine. They published their own work without individual recognition. Why should Graindorge and the other Caennais expect preferential treatment for themselves or for their academy's work? After all, it was the king's science.

The Parisian stance was comprehensible. Unfortunately for all concerned, however, this clash over ownership of Caen's scientific work led to yet another disastrous misunderstanding of the kind that plagued the Académie de Physique's relationship with the royal bureaucracy. Within Graindorge's frame of reference, the Académie de Physique's anatomy program was a disastrous failure. Given the assumptions operative in Paris, it represented a success.

The situation with the Académie de Physique's submissions in technology was much the same. Given the standards operative in Caen, success was marginal. By Parisian standards, however, the academy accomplished a great deal. One of the primary ways Colbert used the Académie Royale des Sciences was as a bureau for technical evaluation.[25] Individual petitioners who claimed to square the circle, solve the longitude problem, or desalinate seawater found themselves forced to submit their projects to evaluation at the Académie Royale before Colbert would even consider rewarding their wondrous discoveries. In every one of the Académie de Physique's technical submissions, the Caen academy contributed to this process, either by generating technical knowledge or by defining technical feasibility.

As with the anatomy program, direct evidence reveals three technical submissions from Caen and strong circumstantial evidence indi-

[24]Tolmer, *Huet*, pp. 385–386.
[25]See Hahn, *Anatomy of a Scientific Institution*, pp. 20–24.

cates a fourth. The first was Chasles's conversion tables for commonly used systems of weights and measures. The second was Jacques Graindorge's "solution" to the longitude problem. The third and fourth submissions came from Hauton's desalinization apparatus and Villons's marine chronometer. All the technical problems addressed by these projects had currency during the 1660s. With such projects even the academy's failures became important. This was particularly the case with both Chasles's project rationalizing weights and measures and Jacques Graindorge's solution to the longitude problem. In those cases submission to Paris destroyed the assumptions that had prompted the work. With the monk, that was the end of the project; the ever-patient Chasles turned to a new approach and went back to work.

With Hauton's desalinization project and Villons's clock things went differently. Both projects were hailed as successes in Paris, whereas these men's achievements went largely unrecognized at home in Caen. In Hauton's case, his trip to Paris in late 1669 may have come at the wrong time. He presented his work to the Académie Royale during a period when Graindorge thought the academy was in its "death shroud." According to Graindorge's report to Huet, most of the other academicians considered Hauton's project dull and inconsequential. Some may have thought it embarrassing. Graindorge defended it for what it was, a practical solution to a real problem; but even he appears to have thought it was only modestly successful. He certainly did not see the project as justifying the existence of a royal academy in Caen.[26]

The same division in reactions occurred with Villons's marine chronometer. The Caennais saw it as a strange and curious clock that never worked quite right; the Parisians saw a technological breakthrough. As with Hauton, the two years Villons needed to perfect his device frustrated Graindorge.[27] Moreover, since Villons worked in secret and then failed to return to Caen following his presentation in Paris, Graindorge was unsure whether the project even reflected any

[26]BL, A 1866, 593, 595 (7 November and 20 December 1669): "Quelqueun ne manqua pas de dire hier a notre Academie que ce nestoit pas un grand secret a porter a paris que de distiller de leau et quil ny avoit pas la rien de nouveau. Je lui repartis sec quon nestoit pas a paris pour dire quon avoit des secrets mais seulement pour fournir de leau potable dans les voyages de long cours . . . que cela soit vieux ou nouveau secret ou public ce nestoit pas la question."

[27]Villons had definitely begun work on the device before January 1669; his presentation in Paris occurred in late 1670 (ibid., 698, 702 [n.d.]).

credit on the Académie de Physique.[28] Certainly, by the time Villons was ready to make his presentation at the Académie Royale, the enthusiasm in Paris came as an anticlimax in Caen. For Graindorge personally, Villons's triumph at the Académie Royale, his commission to build a clock for the king, and his 7,000-livre royal post all meant only that he would no longer have to face wrangling with this long-time adversary.[29]

On balance, the theme uniting the Académie de Physique's technical submissions to Paris is very simple. These were sophisticated attempts to solve technical problems that were engaging scientific minds during the 1660s and 1670s. All fell within the boundaries of the mission Colbert had defined for French royal science. Moreover, the Parisians seem to have discounted the particular problems encountered by each of these projects; on the whole, their reactions were extremely positive.

Overall, in fact, the evidence on how the Parisians saw both the technical and the anatomy submissions from Caen argues that those "gentlemen of Paris" whom Graindorge so often vilified actually sought to encourage their Caennais brethren. From a modern vantage point, the Parisian attitudes are more easily comprehensible than the Caennais position. Clearly the Caennais conceived a modern organizational form for their science. They established an organization devoted to cooperative, systematic research. They demonstrated the skill necessary to produce notable results. They attracted attention to their work, and even earned praise, encouragement, and rewards. Yet they simply did not sustain what they had created. They chose to close their own academy. It is in the dichotomy between their successes and that choice that we find the significance of their story.

[28]Graindorge may have been correct. Chamillart's recommendation of Villons's project to Colbert failed to mention the academy in Caen: "Je n'ai point eu reponse de vostre part sur la fabrique de la montre du sieur chevalier de Villons don i'ai eu l'honneur de vous envoier le device. Si l'accademie Royale en faict estime il se donnera l'honneur de vous la puter. C'est un fort honneste garçon qui a beaucoup de genie pour les machines" (Chamillart to Colbert, 28 February 1669, BN, Mélanges Colbert, 149 bis, f. 661).

[29]BL, A 1866, 702, 655 [September or October 1670, late 1670].

8

Conclusion

The significance of the Académie de Physique's history lies in the disparity between its expressed rationale (its stated organizational mission) and the reality of its scientific practices. Despite early successes as a private-patronage organization, the academy never managed to justify its existence as a state-funded research institute. The disjunction we find on this point suggests a great deal about seventeenth-century French science and the process of creating new organizational forms. More specifically, the academy's history speaks directly to the matrix of historical interpretations that constitute the Fontenelle thesis.

The Académie de Physique underwent a meteoric rise—from informal *assemblée* in 1662 to private society in 1666 to royal academy in 1667. Yet once it was formalized as a private society, its scientific practices never really changed again. The academy's history thus violates the model that the Fontenelle thesis lays down for French science in two important ways. Before 1666 there was no patronage crisis. After 1666 no meaningful changes occurred in the type of scientific work done at the academy, despite the increasing complexity of its organizational structure. For the academy, no causal linkages mandated that more sophisticated (or expensive) scientific practices must create a more complex organizational structure. Exactly the opposite occurred: an increasingly complex and expensive organizational structure demanded justification by more sophisticated science.

Few things obscure our vision of the past so much as success. Certainly this is the case with the history of seventeenth-century French science, the Académie Royale, and state involvement in science. The Fontenelle thesis was built on the assertion that state involvement saved the French scientific community of the 1660s. Yet no historian has ever demonstrated that a patronage crisis actually existed. The only evidence adduced to support the notion has been the assertion that all significant private-patronage groups disappeared from Paris before mid-1665. After all, the successes of the Académie Royale during the eighteenth century made such a failure self-evident—private patronage *must* have failed. Looking only at the post-1699 successes of the Académie Royale, historians since Fontenelle have systematically overlooked the possibility that the "patronage crisis" of the 1660s was chimerical. Only by looking outside Paris can we see the possibility of a different perspective: in the Académie de Physique's history, we find the opening of the Académie Royale actually disrupting private-patronage support for French science.

The Académie de Physique reached the apogee of its working potential in the early months of 1666, while Graindorge acted as intelligencer orchestrating weekly exchanges between the Thévenot and Huet's group. The academy first formalized its existence on the basis of the exchanges with the Thévenot—exchanges that began only in mid-1665. Of course, the coming of the Académie Royale des Sciences destroyed the possibility of continuing exchanges with the Thévenot, and there is simply no way to measure how productive the Académie de Physique could have become if the Thévenot had continued to exist beyond March 1666. Nevertheless, comparisons of the 1665–1666 exchanges with the Thévenot and the Académie de Physique's difficulties in its later dealings with the Académie Royale make it impossible to conceive that the creation of state-controlled institutions contributed to any immediate amelioration of French science.

Concentration on the Académie Royale and the a priori dismissal of all other organizational forms as meaningless after 1663 have seriously distorted our understanding of both the social and intellectual history of French science. As we saw in Chapter 2, for example, only by overlooking the existence of the Thévenot during 1665 and 1666 can historians of the Académie Royale claim Mariotte's discovery of the blind spot in the eye as a triumph for royal science. Far from representing the royal academy's earliest experimental triumph,

Mariotte's "discovery" actually represented nothing more than a reawakening of interest in a phenomenon discussed at the Thévenot over a year before the Académie Royale held its first session.

The existence of the Thévenot and its program certainly undercuts the basic premises of the Fontenelle thesis, but it does little to offer any constructive alternatives for understanding French science. On that issue the Académie de Physique's history offers us a great deal more. After all, Colbert ordered the academy to coordinate its program with the royal institution in Paris. Here we find the great irony to the Académie de Physique's history.

Under the terms Colbert's ministry dictated for the relationship between Paris and Caen, the Académie de Physique had little chance of ever establishing an effective program. Just as Graindorge claimed, the Académie de Physique was at its best when it was allowed to follow the *curiosités* of the new empirical science wherever they might lead. It was at its worst when its membership tried to formulate a specific set of projects to please Chamillart and Colbert. Although theoretically conceivable, an effective working relationship between the royal academies in Paris and Caen proved impossible to effect. Ironically, nothing argues more strongly for that conclusion than the fact that such a successful relationship had once existed between the private-patronage Thévenot and the private-patronage Académie de Physique.

A great deal of evidence suggests that private-patronage science remained a viable organizational form in French science right up to the moment the Académie Royale opened. Moreover, the effort to transform a private-patronage organization into a state institution proved unsuccessful. When we put these two points together, we see that there was no logical or natural organizational progression from private-patronage to state-controlled science during the 1660s. Indeed, the Académie de Physique's history argues that the effort to create royal scientific institutions amounted to a serious disruption of French science.

Bibliography

To place the Académie de Physique within the proper historical context, one must look to works on local history of Caen. Among the sources of that literature, the *Mémoires de l'Académie Nationale des Sciences, Arts, et Belles-Lettres de Caen* contains numerous synthetic articles and monographs that convey historical insights of very high quality. Jean Yver's work on the government of Caen during the seventeenth century (1934) is particularly notable in this regard, as is the work of Jacques Denis on seventeenth-century skepticism (1872), René Delorme's biography of Moisant de Brieux (1872), and M. Gournay's summary biography of Huet (1855).

Going beyond the scope of those works, Katherine Stern Brennan's dissertation, "Culture and Dependencies," offers a broad survey of intellectual life in seventeenth-century Caen. Even broader treatments are Gabriel Vanel's three-volume study and A. Galland's *Histoire du Protestantisme,* M. G. Mancel's edition of an anonymous seventeenth-century diary, and Huet's own *Origines de la ville de Caen.* With its emphasis on Protestantism and the revocation of the Edict of Nantes, Galland's work contains a great deal of general background on economic conditions in Normandy. Huet's *Origines* still ranks as an indispensable source. Jean-Claude Perrot's *Genèse d'une ville moderne* deserves mention as an *Annaliste*'s treatment of early-modern Caen, but Perrot's focus is the eighteenth century. For seventeenth-century background he relies heavily on such traditional sources as Huet's work.

For literature on the Académie de Physique itself, the works of Harcourt Brown, Leon Tolmer, and Katherine Stern Brennan offer the best available treatments. Brown devotes the major portion of a chapter of his *Scientific Organizations* to the Académie de Physique, but the work suffers from an overreliance on Huet's *Commentarius.* Brown later rectified that shortcoming with an extensive article, "L'Académie de Physique," in the *Mémoires* of the academy in Caen (1939). Here he identifies 166 of André Graindorge's letters in the collection at the Laurenziana in Florence and bases his account on those letters and on Huet's *Commentarius* and *Origines.* Although this article contains excellent material, it is limited in scope because Brown did not have access to any Graindorge letters from before late 1665. Several errors in his dating of Graindorge's letters (errors that could be corrected through internal evidence) also flaw this work. In particular, by ascribing numerous letters from late 1666 to 1669, Brown skews the coherence of Graindorge's account and draws an overly dark picture of the academy's activities in 1669.

Leon Tolmer's publication of twenty-two additional Graindorge letters (1943) and his later biography of Huet add considerably to the scholarship on the Académie de Physique. Tolmer's research and his knowledge of Huet are impressive, but his work tends toward antiquarianism and hasty conclusions. Brennan's chapter on the Académie de Physique is more balanced, but since her real focus is the Grand Cheval, her treatment of the Académie de Physique is primarily a synthesis of work already done by Brown and Tolmer. The value of Brennan's work lies in the way she ties the Académie de Physique to the intellectual community in Caen and to the Republic of Letters.

Besides the core of material drawn primarily from local histories, the second area of literature on which this history of the Académie de Physique rests comes from the history of science. For this research, the multitude of short research notes in such journals as the *Revue d'Histoire des Sciences et de Leurs Applications* have proved of far greater value than the more synthetic material found in longer articles, monographs, and general surveys of seventeenth-century science. Here the focus of research on science has been narrow-gauge, some might say antiquarian, in its attempt to lay bare the precise state of practice and the conception of ideas. Such analysis is critical to an assessment of the functional significance of a scientific organization. Among the more general works that contribute to an assessment of seventeenth-century science are such articles as Léon Auger's "R. P. Mersenne et la physique," Charles Bedel's "Pharmacie au XVIIe siècle," Marie Boas Hall's "Establishment of the Mechanical Philosophy," Harcourt Brown's "Utilitarian Motive in the Age of Descartes," Joseph Schiller's "Laboratoires d'anatomie et de botanique à l'Académie des Sciences au XVIIe siècle." Among the more general works in the history of science, Abraham Wolf's *History of Science, Technology, and Philosophy in the 16th and 17th Centuries* and Maurice Daumas's *Instruments scientifiques aux XVIIe et XVIIIe siècles* offer the most useful texts.

For research on such topics as the Académie de Physique, broader surveys are needed in late seventeenth-century French history. On this subject, the basic bibliographic and historiographic guide is William F. Church's *Louis XIV in Historical Thought,* and Ernest Lavisse's *Louis XIV* remains the starting point for any scholarship. Works that deserve special notice are John B. Wolf's *Louis XIV,* which offers a valuable survey and synthesis covering political, economic, and social issues in the second half of the seventeenth century. David Maland's *Culture and Society in Seventeenth-Century France* offers similar scope on high culture, patronage, royal academies, and the Republic of Letters. James King's *Science and Rationalism,* a work too often overlooked, provides a suggestive thesis for research on the political significance of high culture.

Many other works also treat general problems in seventeenth-century French history in ways that bear on this book. Among the more important are the various works of the *Annaliste* Robert Mandrou, Lionel Rothkrug's *Opposition to Louis XIV,* P. W. Fox's "Louis XIV and the Theories of Absolutism and Divine Right," E. H. Kossman's history of the Fronde, R. B. Grassby's "Social Status and Commercial Enterprise under Louis XIV," the essays in John Rule's edited volume *Louis XIV and the Craft of Kingship,* and the entire corpus of Roland Mousnier's work.

On more specialized issues in French history, F. Olivier-Martin's *Organization corporative de la France de l'ancien régime* furnishes the starting point for any discussion of corporatism. And if Olivier-Martin wrote the classic, Emile Coornaert's *Corporations en France avant 1789* is nearly as significant. The works of E. Lousse and D. Richet are also important for their contributions to this topic. Roland Mousnier's writings, particularly his landmark article "Les Concepts d' 'ordres,' d' 'états,' de 'fidèlité' et de 'monarchie absolue' en

France de la fin du XVe siècle à la fin du XVIIIe" (1972), are essential for their definitions of life in a society of orders. William H. Sewell's "Etat, Corps, and Order: Some Notes on the Social Vocabulary of the French Old Regime" offers similar insights. A. L. Moote's *Revolt of the Judges: The Parlement of Paris and the Fronde, 1643–1652* offers extensive treatment of the politics of corporate France. Taking in a longer scope of French history, J. H. Shennan's *Parlement of Paris* also offers insights into corporate society while analyzing the legal foundations of corporatism. Pierre Lefebre's "Aspects de la 'fidèlité' . . ." is a detailed case study of the workings of the society of orders. Joseph Klaits's *Printed Propaganda under Louis XIV* treats royal attempts to control public opinion, as does Peter Fraser's *Intelligence of Secretaries of State.* The best work on the administrative procedures of Louis XIV's government is Julian Dent's *Crisis in Finance.* L. Bernard's *Emerging City* and Orest Ranum's *Paris in the Age of Absolutism* give essential material on Paris and the development of royal policy.

Two recent biographies of Colbert have added significantly to a specialized field that had been too quiet for a good many years. As a descendant of Colbert, Inès Murat gained access to family archives that contain important new materials. Jean Meyer's *Colbert* is an excellent new biography. Pierre Deyon's *Mercantilisme* presents the best *Annales* school treatment of Colbertism. The exhibition catalogue *Colbert, 1619–1683,* published to accompany a 1983 Paris exhibition commemorating the three hundredth anniversary of Colbert's death, contains numerous articles summarizing Colbert scholarship and outlining sources. Pierre Clément's scholarship still remains a standard, both in his editions of Colbert papers and in his analysis. Lavisse's treatment of Colbert, although largely repudiated in its specifics, still remains a suggestive thesis worthy of attention.

For French intellectual life L. W. B. Brockliss's *French Higher Education in the Seventeenth and Eighteenth Centuries* is a recent work that details the spectrum of intellectual pursuits followed in French universities. Numerous older histories offer syntheses of scientific ideas with broader intellectual life. The classic among these works is Paul Hazard's *Crise de la conscience européenne;* Paul Barrière's *Vie intellectuelle* covers similar ground. Gaston Bachelard's *Formation de l'esprit scientifique* focuses more narrowly on the psychology of scientific thought. René Pintard's *Libertinage érudit* is central to efforts to deal with early seventeenth-century intellectual history. Likewise, Léon Wencelin's article "La Querrelle des anciens et de moderns et l'humanisme" gives context to seventeenth-century intellectual life. On patronage and the Republic of Letters, Orest Ranum's *Artisans of Glory* defines Louis XIV's political policies as a major force in the professionalization of intellectual activity. An article by Charles Bioche, "Les Savant et le gouvernement au XVIIe et XVIIIe siècles," makes a similar point very well. Robert Isherwood's *Music in the Service of the King* offers an important case study in royal patronage. Howard Solomon, in his *Public Welfare, Science, and Propaganda in Seventeenth-Century France,* treats politics, patronage, corporatism, and royal propaganda in the context of Theophraste Renaudot's lifework.

On the Académie Royale's history, Roger Hahn's *Anatomy of a Scientific Institution* offers the best starting point. Hahn's work focuses primarily on the eighteenth century, but his chapter on the academy's founding offers an excellent summary of the Fontenelle thesis. Hahn's emphasis on the professionalism of the early academy is perhaps overstated, but he is right to emphasize the self-consciousness that advocates of the academy brought to their efforts. The extensive bibliographical materials also make this work the best guide for research on the Académie Royale. Other works that warrant special attention are Harcourt Brown's *Scientific Organizations in Seventeenth-Century France,* John Milton Hirschfield's *Académie Royale,* and Alfred Joseph George's article "The Genesis of the Academy of Sciences." Brown's work is particularly important for placing the Académie Royale in the larger context of French science. Hirschfield provides the most detailed synthesis describing the academy's origins in terms of the Fontenelle thesis. George's work is highly suggestive on the academy's corporate context.

Too little comparative analysis has been done on institutional and organizational developments in French and English science. Besides Martha Ornstein's classic *Rôle of Scientific Societies in the Seventeenth Century,* most of that work has been sociological rather than historical in focus. Examples can be found in the work of Joseph Ben-David and Wolfgan Van Den Daele. Of course, a vast literature exists on the Royal Society, and unlike the French case, the society has generated heated controversy over its origins. Some of the more important recent works include Michael Hunter's *Science and Society in Restoration England,* Charles Webster's *Great Instauration,* and various works by Margaret and James Jacob. Hunter's work provides an important statement on the need to place Restoration science (including the

Royal Society) in larger political, religious, and philosophical frameworks. Webster's work is an exhaustive survey of the institutional basis of English science in the seventeenth century, but it suffers from a narrow vision of English society. The Jacobses' attempt to tie English science to specific theological positions is suggestive but highly controversial.

As part of the debates over the Royal Society's early history, numerous scholars have expanded their focus to treat English science within a larger intellectual framework. On the whole, this appears to be an extremely productive shift. Notable examples of this trend include Michael Hunter's work, Mordechai Feingold's *Mathematician's Apprenticeship,* Harold J. Cook's *Decline of the Old Medical Regime in Stuart London,* and Barbara Shapiro's *Probability and Certainty.* Feingold treats science in the English universities in the late sixteenth and early seventeenth centuries. Cook uses professional politics to explain the shifting scientific interests of medical practitioners over the seventeenth century. Shapiro provides a truly synthetic survey of intellectual life in seventeenth-century England.

A Note on Archival Sources

The major archival sources used in the preparation of this book can be found in collections housed at the Biblioteca Medicea-Laurenziana, in Florence, and at the Bibliothèque Nationale and the Académie des Sciences, in Paris. Additional papers and collections were consulted in the Municipal Archives of Caen, in the Departmental Archives of Calvados, and in the Archives Nationales, Paris.

The essential body of documents revealing the history of the Académie de Physique is found in the 191 letters André Graindorge addressed to Pierre-Daniel Huet between 1661 and 1675. Of these letters, 169 are part of the Ashburnham 1866 collection housed in the Biblioteca Medicaea-Laurenziana. (Readers interested in the scope of the collection of Huet papers housed in Florence will find the works of Léon G. Pellissier listed below extremely useful. Those interested in the history of the collection should also consult the bibliographic entry under C. P. Tollet.) Graindorge's remaining twenty-two letters were published by Léon Tolmer in the *Mémoires* of the modern academy in Caen, listed below. In addition to the Graindorge letters, Ashburnham 1866 (the Huet papers) contains numerous other documents related to the Académie de Physique.

In Paris, numerous documents housed in the manuscript collections at the Bibliothèque Nationale relate to the Académie de Physique. Among the most important are documents among the Huet papers: Fonds Français, 11903–11914, 14555–14558, 15189, 15351, 22613; Fonds Français, Nouvelles Acquisitions, 1174, 1197, 5856, 6202; and Fonds Latin, 11453, 15188–15190. Other materials can be found in Baluze, 325, and throughout the Colbert collections known as the Mélanges de Colbert and the Cinq Cents de Colbert. Also in Paris, additional material on the Académie de Physique is found at the Institut de France, Archives de l'Académie des Sciences, "Registre des procès verbaux des séances."

Selected Works

Aucoc, Léon, ed. *L'Institut de France: Lois, status et règlements concernant les anciennes académies et l'Institut de 1635 à 1889*. Paris, 1889.
Auger, Léon. "Le R. P. Mersenne et la physique." *RHS* 2 (1948): 33–52.
Avenal, J. d'. *Histoire de la vie et des ouvrages de Daniel Huet, evèque d'Avranches*. Mortain, 1853.
Bachelard, Gaston. *La Formation de l'esprit scientifique*. Paris, 1947.
Barrière, P. *La Vie intellectuelle en France du XIVe siècle à l'époque contemporaine*. Paris, 1962.
Barthomèss, Christian. *Huet évêque d'Avranches, ou Le Scepticisme théologique*. Paris, 1850.
Bedel, Charles. "La Pharmacie au XVIIe siècle." *XVIIe siècle*, no. 30 (1956), pp. 46–61.
Belloni, Luigi. "Marcello Malpighi and the Founding of Anatomical Microscopy." In *Reason, Experiment, and Mysticism in the Scientific Revolution*, ed. M. L. Righini Bonelli and William R. Shea. New York, 1975.
Ben-David, Joseph. "Scientific Growth: A Sociological View." *Minerva* 2 (1964): 455–476.
——. "The Scientific Role: The Conditions of Its Establishment in Europe." *Minerva* 4 (1965): 15–54.
Bernard, L. *The Emerging City: Paris in the Age of Louis XIV*. Durham, N.C., 1970.
Bertrand, J. *L'Académie des Sciences et les académiciens de 1666–1795*. Paris, 1869.
Bigourdan, G. "L'Académie de Montmor." *Comptes Rendus de l'Académie des Sciences* 164 (1917): 217–218.
——. *L'Astronomie: Évolution des idées et des méthodes*. Paris, 1911.
——. *Les Premières Sociétés savants de Paris et les origines de l'Académie des Sciences*. Paris, 1919.
Bioche, Charles. "Les Savants et le gouvernement au XVIIe et XVIIIe siècles." *Revue de Deux Mondes* 107 (1937): 174–189.
Boas, Marie. "Boyle as a Theoretical Scientist." *Isis* 41 (1950): 261–268.
——. "The Establishment of the Mechanical Philosophy." *Osiris* 10 (1952): 412–541.

——. "Quelques Aspects sociaux de la chimie au XVIIe siècle." *RHS* 10 (1957): 132–147.

Bochart, Samuel. *Opera omnia*. Leyden, 1712.

Bonelli, M. L. Righini, and William R. Shea, eds. *Reason, Experiment, and Mysticism in the Scientific Revolution*. New York, 1975.

Brennan, Katherine Stern. "Culture and Dependencies: The Society of the Men of Letters of Caen from 1652 to 1705." Ph.D. dissertation, Johns Hopkins University, 1981.

Brockliss, L. W. B. *French Higher Education in the Seventeenth and Eighteenth Centuries: A Cultural History*. Oxford, 1987.

——. "Medical Teaching at the University of Paris, 1600–1720." *AS* 35 (1978): 221–251.

Brown, Harcourt. "L'Académie de Physique de Caen (1666–1675) d'après les lettres d'André de Graindorge." *MANC*, n.s. 9 (1939): 117–208.

——. *Scientific Organizations in Seventeenth-Century France*. Baltimore, 1934.

——. "The Utilitarian Motive in the Age of Descartes." *AS* 1 (1936): 182–192.

Brugmans, Henri L. *Le Séjour de Christian Huygens à Paris*. Paris, 1935.

Bugler, G. "Un Précurseur de la biologie expérimentale: Edme Mariotte." *RHS* 3 (1950): 242–250.

Caullery, Maurice. "La Biologie au XVIIe siècle." *XVIIe Siècle*, no. 30 (1956), pp. 25–45.

Church, William F. *Louis XIV in Historical Thought*. New York, 1976.

Clément, Pierre. *Histoire de la vie et l'administration de Colbert*. Paris, 1846.

——, ed. *Lettres, instructions et mémoires de Colbert*. 7 vols. Paris, 1861–1882.

Colbert, 1619–1683. Paris: Ministère de la Culture, 1983.

Cole, C. W. *Colbert and a Century of French Mercantilism*. 2 vols. New York, 1939.

Collas, G. *Jean Chapelain*. Paris, 1911.

Collier, Katherine B. *Cosmogonies of Our Fathers*. New York, 1968.

Cook, Harold J. *The Decline of the Old Medical Regime in Stuart London*. Ithaca, N.Y., 1986.

Coornaert, Emile. *Les Corporations en France avant 1789*. Paris, 1941.

Daumas, Maurice. *Les Instruments scientifiques aux XVIIe et XVIIIe siècles*. Paris, 1953.

——. "Le Mythe de la révolution technique." *RHS* 16 (1963): 289–302.

——. "La Vie scientifique au XVIIe siècle." *XVIIe Siècle*, no. 30 (1956), pp. 110–132.

Delorme, René. "Moisant de Brieux, fondateur de l'académie de Caen." *MANC*, 1872, pp. 27–110.

Denis, Jacques. "Sceptiques et libertins de la première moitier du XVIIe siècle." *MANC*, 1884, pp. 184–201.

Dent, Julian. *Crisis in Finance: Crown, Financiers, and Society in Seventeenth-Century France*. New York, 1973.

Depping, G., ed. *Correspondance administrative sous le règne de Louis XIV*. 4 vols. Paris, 1850–1855.

Deyon, P. *Le Mercantilisme*. Paris, 1969.

Dionis, Pierre. *L'Anatomie de l'homme suivant la circulation du sang*. Paris, 1968.

Duchesneau, François. "Malpighi, Descartes, and the Epistemological Problems of Iatromechanism." In *Reason, Experiment, and Mysticism in the Scientific Revolution*, ed. M. L. Righini Bonelli and William R. Shea. New York, 1975.

Dulieu, Louis. "La Contribution montpelliéraine aux recueils de l'Académie Royale des Sciences." *RHS* 11 (1958): 250–262.

——. "Le Mouvement scientifique montpelliéraine au XVIIIe siècle." *RHS* 11 (1958): 227–249.

Dupront, A. *P.-D. Huet et l'exégèse comparatiste au XVIIe siècle.* Paris, 1930.

Faure-Fremiet, E. "Les Origines de l'Académie des Sciences de Paris." *Notes and Records of the Royal Society, London* 21 (1966): 20–31.

Feingold, Mordechai. *The Mathematicians' Apprenticeship: Science, Universities, and Society in England, 1560–1640.* Cambridge, Eng., 1984.

Feron, A. *Notes sur les académies provinciales.* Rouen, 1934.

Formigny de La Londe, A.-R. R. de. *Documents inédits pour servir à l'histoire de l'ancienne Académie Royale des Belles-Lettres de Caen.* Caen, 1854.

Fox, P. W. "Louis XIV and the Theories of Absolutism and Divine Right." *Canadian Journal of Economics and Political Science* 26 (1960): 128–142.

Fraser, Peter. *The Intelligence of Secretaries of State and Their Monopoly of Licensed News, 1660–1688.* Cambridge, Eng., 1956.

Fuller, Lon. *Legal Fictions.* Stanford, Calif., 1967.

Galland, A. *L'Histoire du Protestantisme à Caen et en Basse-Normandie de l'Edit de Nantes à la Révolution.* Paris, 1898.

Gauja, Pierre. "L'Académie Royale des Sciences (1666–1793)." *RHS* 1 (1949): 293–310.

——. "Les Origines de l'Académie des Sciences de Paris." Institut de France, Académie des Sciences, *Troisième Centenaire, 1666–1966,* pp. 1–51. Paris, 1967.

George, Albert Joseph. "The Genesis of the Academy of Sciences." *AS* 3 (1938): 372–401.

——. "A Seventeenth-Century Amateur of Science, Jean Chapelain." *AS* 3 (1938): 217–236.

Godard, Charles. *Les Pouvoirs des intendants sous Louis XIV, particulièrement dans les pays d'élections, de 1661 à 1715.* Geneva, 1974 [1901].

Gosselin, D. "Origines et petite histoire de l'Académie de Caen à parti d'une pièce de vers inédite." *MANC,* n.s. 17 (1970): 215–228.

Gouhier, P. "Autour des origines de l'Académie de Caen." *Bulletin de la Société des Antiquaires de Normandie* 58 (1965–1966): 415–418.

Gournay M. "Huet évêque d'Avranches." *MANC,* 1855, pp. 318–396.

Grassby, R. B. "Social Status and Commercial Enterprise under Louis XIV." *Economic History Review,* 2d ser., 13 (1960): 1938.

Guiffrey, J. J. *Comptes des Bâtiments du Roi sous le règne de Louis XIV.* 5 vols. Paris, 1891–1901.

Hahn, Roger. *The Anatomy of a Scientific Institution: The Paris Academy of Sciences, 1666–1803.* Berkeley, 1971.

Hall, A. Rupert, and Marie Boas Hall, eds. and trans. *The Correspondence of Henry Oldenburg.* 9 vols. Madison, Wis., 1965–1973.

Hall, Marie Boas. "Salomon's House Emergent: The Early Royal Society and Cooperative Research." In *The Analytic Spirit,* ed. Harry Woolf. Ithaca, N.Y., 1981.

——. "Science in the Early Royal Society." In *The Emergence of Science in Western Europe,* ed. Maurice Crosland, pp. 57–77. New York, 1976.

Hartung, F., and R. Mousnier, "Quelques Problèmes concernant la monarchie absolue." In *Relazioni de X Congresso Internazionale de Scienze Storiche, Rome, 1955,* 4:3–55. Florence, 1955.

Hazard, Paul. *La Crise de la conscience européenne.* Paris, 1935.

Henry, Charles. "Pierre de Carcavy, intermédiare de Fermat, de Pascal et de Huygens, bibliothècaire de Colbert et du Roi, directeur de l'Académie des Sciences." *Bulletino Boncampagni,* 1884, pp. 317–391.

Huard, Georges. *Deux Académiciens caennais des XVIIe et XVIIIe siècles: Les Croismare, seigneurs de Lasson.* Caen, 1921.

Huet, Pierre-Daniel. *Les Origines de la ville de Caen, et des lieux circonvoisins.* 2d ed. Rouen, 1706.

——. *Petri Danielis Huetii Commentarius de rebus ad illum pertinentibus.* Amsterdam, 1718.

Humbert, Pierre. "Mersenne et les astronomes de son temps." *RHS* 11 (1948): 29–32.

Hunter, Michael. "The Debate over Science." In *The Restored Monarchy,* ed. J. R. Jones. London, 1979.

——. *Science and Society in Restoration England.* Cambridge, 1981.

Isherwood, Robert M. *Music in the Service of the King.* Ithaca, N.Y., 1973.

Jacob, James R., and Margaret Jacob. "The Anglican Origins of Modern Science." *Isis* 71 (1980): 251–267.

Johnson, Francis R. "Gresham College: Precursor of the Royal Society." *Journal of the History of Ideas* 1 (1940): 413–438.

Jones, H. W. "La Société Royale de Londres au XVIIe siècle: Reflexions diverses." *RHS* 12 (1950): 214–221.

Kettering, Sharon. *Patrons, Brokers, and Clients in Seventeenth-Century France.* New York, 1986.

King, James E. *Science and Rationalism in the Age of Louis XIV, 1661–1683.* Baltimore, 1949.

Klaits, Joseph. *Printed Propaganda under Louis XIV.* Princeton, 1976.

Kossman, E. H. *La Fronde.* Leiden, 1954.

Lavisse, Ernest. *Louis XIV.* 2 vols. Paris, 1978 [1903–1911].

Lecanu, M. "Etude historique et littéraire sur Pierre-Daniel Huet, évêque d'Avranches." *Journal des Savants de Normandie,* 1844, pp. 515–538, 561–590.

Lefebre, Pierre. "Aspects de la 'fidèlité' en France au XVIIe siècle: Le Cas des agents des princes de Condé." *Revue Historique* 250 (1973): 59–106.

Lenoble, Robert. *Esquisse d'une histoire de l'idée de nature.* Paris, 1968.

——. *Mersenne, ou La Naissance du mécanisme.* Paris, 1943.

——. "Quelques Aspects d'une révolution scientifique à propos du troisième centenaire de P. Mersenne." *RHS* 2 (1948): 53–79.

——. "La Représentation du monde physique à l'époque classique." *XVIIe Siècle,* no. 30 (1956), pp. 5–24.

Lenski, Gerhard. *A Theory of Social Stratification.* Chapel Hill, N.C., 1966.

Lousse, E. "La France d'ancien régime, État corporatif." *Annales de Droit et des Sciences Politiques,* 1937, p. 690.

——. *La Sociéte d'ancien régime: Organization et représentation corporatives.* Louvain, 1943.

Lux, David S. "Royal Patronage and Seventeenth-Century Science: L'Académie de Physique de Caen, 1662–1672." Ph.D. dissertation, University of Michigan, 1983.
McKeon, Robert M. "Une Lettre de Melchisédech Thévenot sur les débats de l'Académie Royale des Sciences." *RHS* 18 (1965): 1–6.
Maindrou, Ernest. *L'Académie des Sciences.* Paris, 1888.
Maland, David. *Culture and Society in Seventeenth-Century France.* London, 1970.
Mancel, M. G. *Journal d'un bourgeois de Caen (1652–1733).* Caen, 1848.
Mandrou, Robert. *La France aux XVIIe et XVIIIe siècles.* Paris, 1967.
——. *From Humanism to Science.* Trans. Brian Pearce. New York, 1978.
——. *Introduction à la France moderne: Essai de psychologie historique, 1500–1640.* Paris, 1961.
——. *Louis XIV en son temps, 1661–1715.* Paris, 1973.
Masson, G. *Queen Christina.* New York, 1968.
Mesnard, Jean. "Pascal à l'Académie le Pailleur." *RHS* 16 (1963): 1–10.
Metzger, Hélène. *Les Doctrines chimiques en France du début du XVIIe à la fin du XVIIIe siècle.* Paris, 1923.
Meyer, J. *Colbert.* Paris, 1981.
Middleton, W. E. Knowles. *The Experimenters: A Study of the Accademia del Cimento.* Baltimore, 1971.
——. *The History of the Barometer.* Baltimore, 1964.
Millepierres, François. *La Vie quotidienne des médecins au temps de Molière.* Paris, 1964.
Millington, E. C. "Theories of Cohesion in the Seventeenth Century." *AS* 5 (1945): 253–269.
Mongrédien, G. *Colbert.* Paris, 1963.
Moote, A. L. *The Revolt of the Judges: The Parlement of Paris and the Fronde, 1643–1652.* Princeton, 1971.
Mousnier, Roland. "Les Concepts d' 'ordres,' d' 'états,' de 'fidélité' et de 'monarchie absolue' en France de la fin du XVe siècle à la fin du XVIIIe." *Revue Historique* 247 (1972): 289–312.
——. *État et société en France aux XVIIe et XVIIIe siècles.* Paris, 1969.
——. "L'Évolution des Institutions monarchiques en France et ses relations avec l'état social." *XVIIe Siècle,* nos. 58–59 (1963), pp. 57–72.
——. *Les Hiérarchies sociales de 1450 à nos jours.* Paris, 1969.
——. *Les Institutions de la France sous la monarchie absolue.* Vol. 1. Paris, 1974.
——. *Les XVIe et XVIIe Siècles.* Paris, 1965.
Murat, Inès. *Colbert.* Paris, 1980.
Musset, Lucien. "L'Académie de Caen à travers trois siècles de son histoire." *MANC,* n.s. 17 (1970): 193–202.
Olivier-Martin, F. *L'Organization corporative de la France d'ancien régime.* Paris, 1938.
Ornstein, Martha. *The Rôle of Scientific Societies in the Seventeenth Century.* 2d ed. London, 1963.
Patterson, T. S. "Van Helmont's Ice and Water Experiments." *AS* 1 (1936): 462–467.

Pelissier, Léon G. "À travers les papiers de Huet." *Bulletin du Bibliophile,* 1888, pp. 385–411, 503–529; 1889, pp. 29–59.
——. "Inventaire sommaire des papiers de P.-D. Huet à la Bibliothèque Laurentienne de Florence." *Revue des Bibliothèques* 9 (1899): 1–2; 10 (1900): 67–79.
Perrault, Charles. *Mémoires de ma vie.* Ed. Paul Bonnefon. Paris, 1909.
Pillet, Victor Evremont. "Étude sur Antoine Halley." *MANC,* 1858, pp. 173–224.
Pintard, Réne. *Le Libertinage érudit dans la première moitié du XVIIe siècle.* Paris, 1943.
Popkin, Richard. "The High Road to Pyrrhonism." *American Philosophical Quarterly* 2 (1965): 18–32.
——. *The History of Skepticism from Erasmus to Spinoza.* Berkeley, 1979.
——. "The Manuscript Papers of Pierre-Daniel Huet." In *Year Book of the American Philosophical Society, 1959,* pp. 449–453. Philadelphia, 1959.
——. "The Skeptical Crisis and the Rise of Modern Philosophy." *Review of Metaphysics* 7 (1953–1954): 132–151, 307–322, 499–510.
Quemada, Bernard. *Introduction à l'étude du vocabulaire médical, 1600–1710.* Besançon, 1955.
Ranum, Orest. *Artisans of Glory: Writers and Historical Thought in Seventeenth-Century France.* Chapel Hill, N.C., 1980.
——. *Paris in the Age of Absolutism.* New York, 1968.
Roche, Daniel. *Le siècle des lumières en province: Académies et académiciens provinciaux, 1680–1789.* 2 vols. La Haye, 1978.
Roger, Jacques. "Réflexions sur l'histoire de la biologie (XVIIe–XVIIIe siècle): Problèmes de méthodes." *RHS* 17 (1964): 25–40.
Rothkrug, Lionel. *Opposition to Louis XIV: The Political and Social Origins of the French Enlightenment.* Princeton, 1965.
Rule, J. C., ed. *Louis XIV and the Craft of Kingship.* Columbus, Oh., 1969.
Salomon-Bayet, C. *L'Institution de la Science et l'expérience du vivant: Méthode et expérience à l'Académie Royale des Sciences, 1666–1793.* Paris, 1978.
Schiller, Joseph. "Les Laboratoires d'anatomie et de botanique à l'Académie des Sciences au XVIIe siècle." *RHS* 17 (1964): 97–114.
Sendrail, Marcel. "La Médecine au grand siècle." *XVIIe Siècle,* no. 35 (1957), pp. 163–170.
Sewell, William H. "État, Corps, and Ordre: Some Notes on the Social Vocabulary of the French Old Regime." In *Sozialgeschichte Heute: Festschrift für Hans Rosenberg zum 70. Geburtstag,* ed. Hans-Ulrich Wehler. Göttingen, 1974.
Shapin, Steven, and Simon Schaffer. *Leviathan and the Air-Pump: Hobbes, Boyle, and the Experimental Life.* Princeton, 1985.
Shapiro, Barbara J. *Probability and Certainty in Seventeenth-Century England: A Study of the Relationships between Natural Science, Religion, History, Law, and Literature.* Princeton, 1983.
Shennan, J. H. *The Parlement of Paris.* Ithaca, N.Y., 1968.
Smith, Edouard-Herbert. "Recherches sur la vie et les principaux ouvrages de Samuel Bochart." *MANC,* 1836, pp. 341–377.
Solomon, Howard M. *Public Welfare, Science, and Propaganda in Seventeenth-Century France.* Princeton, 1972.
Tamizey de Larroque, Philippe, ed. *Lettres de Jean Chapelain.* 2 vols. Paris, 1883.
Taton, René. *Les Origines de l'Académie Royale des Sciences.* Paris, 1965.

Tollet, C. P. "Libri, le fameux voleur de livres." *Revue des Bibliothèques,* April–June 1927.

Tolmer, Léon. *Pierre-Daniel Huet, 1630–1721: Humaniste-physicien.* Bayeux, n.d. [1949].

——. "Vingt-deux Lettres inédites d'André Graindorge à P.-D. Huet." *MANC,* n.s. 12 (1942): 245–337.

Vaillé, Eugène. *Histoire générale des postes française.* Vol. 3, *De la réforme de Louis XIII à l surintendance générale des postes (1630–1688).* Paris, 1950.

Van Den Daele, Wolfgang. "The Social Construction of Science: Institutionalization and Definition of Positive Science in the Latter Half of the Seventeenth Century." In *The Social Production of Scientific Knowledge: Sociology of the Sciences,* ed. Everett Mendelsohn, Peter Weingart, and Richard Whitley, 1:27–54. Dordrecht, 1977.

Vanel, Gabriel. *Une Grande Ville au XVIIe et XVIIIe siècles: La Vie publique à Caen et les lieux corconvoisins.* 2d ed. Rouen, 1706.

Webster, Charles. *The Great Instauration.* London, 1973.

Wencelins, Léon. "La Querrelle des anciens et de moderns et l'humanisme." *XVIIe Siècle,* nos. 9–10 (1951), pp. 15–34.

Wiener, P. P. "The Experimental Philosophy of Robert Boyle." *Philosophical Review* 41 (1932): 594–609.

Wolf, Abraham. *A History of Science, Technology, and Philosophy in the 16th and 17th Centuries.* London, 1835.

Wolf, C. *Histoire de l'Observatoire de Paris de sa fondation à 1793.* Paris, 1902.

Wolf, J. B. *Louis XIV.* New York, 1968.

Yver, Jean. "La Ville de Caen: Le Gouverneur et les premiers intendants de 1636 à 1679." *MANC,* n.s. 7 (1934): 313–371.

Index

Académie de France de Rome, 88n, 167
Académie de Peinture et de Sculpture, 55, 88
Acadḿie des Jeux Floraux, 9n
Académie due Grand Cheval, 9n, 15–16, 19–20, 23, 35–36, 43n, 96–97, 156
Académie Française, 11n, 16, 94
Académie Royale des Sciences: and Académie de Physique, 106, 112, 117, 123–125, 128–129, 132, 136–139, 146, 148–150, 175–179; founding of, 3–7, 51–56, 161–162; and private patronage, 51–56; working procedures of, 52, 94
Academy of Inscriptions, 55, 88–89, 166, 169n
Accademia del Cimento, 66, 68, 94
Approbation. See Louis XIV
Assemblée, 8–9, 24–30, 36–38, 98, 114
Auzout, Adrian, 5–6, 20, 52, 56, 58, 166n
Avoye, M., 60n, 65n

Barometer, 69–71, 75–76
Blind spot (optical illusion), 41, 181
Blood: circulation of, 39n, 41, 71–72; transfusion of, 71, 78–79, 136
Bochart, Samuel, 14–15, 18, 68
Boyle, Robert, 77, 83, 123
Brieux, Jacques Moisant de, 15–16, 20, 156
Busnel, Charles, 59, 64–65, 68–69, 85, 97, 103, 126, 162

Cally, Pierre, 90n, 102–103, 113–114, 162
Carcavi, Pierre, 115, 137, 145–146, 152–155, 158–159, 164–168, 176
Cartesianism, 97–98, 103
Chamillart, Guy, 78, 82; as administrator of Académie de Physique, 89–92, 96–98, 114–120, 132–139, 141–148, 154–163; intendance of, 87–88, 95–97
Chapelain, Jean, 15, 18, 31–32, 61, 77
Chasles, Jacques, 90n, 102, 134, 138–139, 141, 158, 162, 170; and weights and measures, 102, 108–110, 115, 126, 128–129, 178
Christina, queen of Sweden, 14–15
Colbert, Jean-Baptiste, 5, 9, 52–56, 95, 105–106, 115–119, 121, 123–124, 129–135, 154–158; and funding for Académie de Physique, 78, 82–83, 115, 123, 133, 136–139, 151–154; and interest in research, 121–122, 173–174, 179; royal administration of, 94, 115–122, 164–173, 182; and royal incorporations, 9, 52–53, 172–174
Constant. See Graindorge, André: empiricism of

Curieux. See Graindorge, André: empiricism of
Curiosité. See Graindorge, André: empiricism of

Daleau (apothecary), 125–126, 135, 138–139, 143, 150, 162, 164
Desalinization of seawater, 63, 90, 107–108, 117, 122, 128, 136, 162, 168
Dissections, 8, 23, 27; of camel, 123; of carp, 42–43; of chicken, 44; of cow, 41; of dogs, 42, 45, 93; of dolphin, 144n; of eye, 40, 46; of fly, 42; of fox shark, 68–69; of frog, 41, 44; of greyhounds, 45; of heart, 40, 68–69; of horse, 40; of human fetus, 126; of macreuse, 176; of mole, 123; of monkfish, 128; of oyster, 123; of porcupine, 123; of rabbit, 39; of screechowl, 80, 90; of snakes, 127; of sparrowhawk, 123; of sturgeon, 124, 154–159, 175–176
—treatises on: *1668,* 112, 126–131, 175; *1670,* 112, 149–150, 153, 175
Du Fresne (artist), 125–126, 128, 175
Du Hamel, Jean-Baptiste, 71, 73, 76, 126

Expérience. See Graindorge, André: empiricism of

Fogel, Martin, 65n
Fontenelle, Bernard de, 1–3
Fontenelle thesis, 4–7, 54–56, 180–182
Fouquet, Nicholas, 55, 88
Fronde, 55, 88

Gallois, Jean, 124, 126, 129, 131, 158, 175–176
Graindorge, André: ambitions of, 12, 18–22; and belles lettres, 19–20, 38–51, 42–44, 50, 154–159; and concept of research, 18–22, 64, 79, 155–159; and definition of scientific practice, 38–51; dissections by, 68–69, 71–73, 75; education of, 10–11; empiricism of, 46–51, 66–71, 73–75, 106–107, 155–159; Paris letters of, 29–36; and *physique* vs. *mathématique,* 47–48, 62, 72–73, 85, 101; social standing of, 10–11, 59–62, 126–131, 142–143; and use of term *académie,* 32, 36–38, 65, 78–80
Graindorge, Jacques, 105–107, 132–135, 178
Graindorge de Prémont, Jacques. *See* Prémont, Jacques Graindorge de
Grande Académie, 52–56
Gratifications. See Royal *gratifications*

Halley, Antoine, 101n
Hauton, Pierre, 62–65, 73, 90n, 100, 134, 147; and desalinization of seawater, 107–108, 117, 122, 128, 136, 162, 168; and royal *gratification,* 63, 107
Huet, Pierre-Daniel: and anatomy, 38–46; and astronomy, 23, 28, 31, 104–105, 145; career of, 9–17, 79; education of, 9, 12–17; the *Origenis Commentaria,* 12–17, 30–32, 42–43, 74, 79, 81; royal pensions of, 11n; and treatise on dew, 73–75
Huygens, Christiaan, 61, 106, 110

Journal des Savants, 61, 69, 86
Justel, Henri, 102, 105, 123–124, 129–130

La Ducquerie, Jean-Baptiste Callard de, 90n, 103, 110, 147, 162
Lantin, Jean-Baptiste, 52n
Lasson, Nicholas Croixmare de, 48–49, 60–62, 65, 90n, 93, 100; and optical instruments, 105, 110–11, 126–127, 162, 167
Longitudes, 105–107, 132–135, 178
Louis XIV: and *approbation* of Académie de Physique, 1–3, 78, 82–83, 87, 115–119, 174; political reforms of, 54–55; and Republic of Letters, 54–56
Lower Normandy, intendance of, 87–88, 95–97

Mariotte, Edme, 41, 181
Montmor Academy, 5n, 161

New Age of Academies. *See* Fontenelle thesis

Oldenburg, Henry, 24, 52–53, 83, 123–124, 129–130, 148

Pascal, Blaise, 70–71
Patronage, functions of, 84, 92–100, 119–120, 133–135, 150–151, 159–163

Patronage relationship (personal), 10–11, 55–56; of Bochart and Huet, 14–15, 18; of Chapelain and Huet, 15, 18, 31–32, 77; of Huet and Graindorge, 10–12, 17–23, 45, 50–51, 58, 73–75, 79, 113–114, 141–144, 153–154, 163
Perrault, Claude, 158, 176
Philosophical Transactions, 86, 101–102, 126
Picard, Jean, 71, 73, 76, 78n, 106, 110
Poisons, 49, 66–68
Postel, Nicolas, 101n, 104, 147, 158, 162, 170
Prémont, Jacques Graindorge de, 11, 18–19
Public works projects, 91–92, 95–97, 99–100, 107, 119, 123, 135, 144

Royal funding, 138–148
Royal *gratifications*, 3, 11, 55, 63, 108, 111–112, 121, 136–137, 166–168
Royal incorporation: of Académie de Physique, 82–84, 87–90, 113–120, 173–174; of Académie Royale des Sciences, 51–56, 89; of Academy of Inscriptions, 88–89; of Montpellier, 51
Royal Society of London, 66, 68

Saint-Aignan, duc de, 169–173
Savary, Jacques, sieur de Courtsigny, 90n, 98n, 101–102, 126, 134; translates *Philosophical Transactions*, 101–102, 126
Scientific instruments, 60–61, 69–71, 90–91, 105, 110–112, 126–127
Sorbière, Samuel, 5–6, 20
Steno, Niels, 39–41, 57, 68, 71

Technology, 95–96, 104–112, 177–179. *See also* Desalinization of seawater; Longitudes; Public works projects; Scientific instruments; Timekeeping; Weights and measures
Thévenot, Melchisédec, 28–30, 56
Thévenot Academy, 4–5, 28–30, 49–51, 65–66; closes, 52, 54–56; and communications with Caen, 32–36, 41, 181–182; scientific program of, 39–42
Timekeeping, 178–179

University (Caen) opposes royal incorporation, 97–98

Vaucouleurs, Matthieu Maheust de, 62–65; and experiments with barometers, 69–71, 75–76, 90–91, 97, 98n, 100, 110, 128, 141, 147, 162
Vavasseur, Pierre le, 101n, 103–104, 162
Vicquemant, Corneille, 58n
Villons, Jean Gosselin, chevalier de, 59–62, 65, 71, 90n, 105, 110–112, 126–127, 135, 144, 148–149, 162; and marine chronometer, 60, 110–112, 149, 153, 168, 178–179
Vossius, Isaac, 14, 72

Weighing air. *See* Barometer
Weights and measures, 108–110, 126, 178

Library of Congress Cataloging-in-Publication Data

Lux. David Stephan.

Patronage and royal science in seventeenth-century France : the Académie de physique in Caen/David S. Lux.

p. cm.

Bibliography: p.

Includes index.

ISBN 0–8014–2334–1 (alk. paper)

1. Académie de physique (Caen. France)—History—17th century.
2. Science—France—Societies, etc.—History—17th century.
3. Science—France—History—17th century. 4. Benefactors—France—History—17th century. I. Title.

Q46.L89 1989

506′.044—dc20 89–1002